FROM IDEA TO SITE

A project guide to creating better landscapes

Claire Thirlwall

RIBA ⬦ Publishing

© RIBA Publishing, 2020

Published by RIBA Publishing, 66 Portland Place, London, W1B 1NT

ISBN 978-1-85946-843-2

British Library Cataloguing-in-Publication Data
A catalogue record for this book is available from the British Library.

Commissioning Editor: Elizabeth Webster
Assistant Editor: Clare Holloway
Production: Richard Blackburn
Designed and typeset by Sara-Miranda Icaza
Printed and bound by Short Run Press Limited, Exeter
Cover image: Photographer Timothy Soar

While every effort has been made to check the accuracy and quality of the information given in this publication, neither the Author nor the Publisher accept any responsibility for the subsequent use of this information, for any errors or omissions that it may contain, or for any misunderstandings arising from it.

www.ribapublishing.com

'Land really is the best art.'

Andy Warhol
(America, 1985)

For Paul – my favourite landscape architect

TABLE OF CONTENTS

THE DESIGN TEAM

CONSTRUCT AND MANAGE

MAINTAIN AND EVALUATE

ACKNOWLEDGEMENTS

My thanks go to; Will Burchnall, Paul F, Professor James Hitchmough, the Maggie's Centre media team, Donncha O'Shea and colleagues at Gustafson Porter + Bowman, and Howard Wood for so generously providing the information and many of the images for the case studies. The case studies make the book – sorry there were quite so many questions.

Martin Brown (who persuaded me this was a good idea), Dr Jenni Barrett, Penny Burt, Jennifer Forakis, Paul Gibbs, Alexander Quartly, Anneli Thomson and Paul Wilkinson for advice, moral support, and distracting me away from the computer when needed.

Stephen Alderton, Andrew Dobson, Kenny Duncan and colleagues at Crest Nicholson, Rob Donald, Mark Farmer, Stewart Gilmour, Lynda Harris, Andrew Kinsey, Vanessa Jukes, Colin Moore, Tom Oulton, Sue Palmer, Jake Robinson, Duncan Read, Keith Rowe, Mike Shilton and Britt Warg (my Swedish translation checker) and those on Twitter, Stackflow and LinkedIn who answered my numerous and sometimes obscure questions with patience and enthusiasm.

Elaine Cresswell, Henry Fenby-Taylor and Ian Lanchbury for reviewing the initial proposal – your comments made an immense difference – and David Jarvis for reading the first draft.

Annemarieke de Bruin, Michael Clinton, Anna Dekker, Catherine Eldred, Professor Paul Ekins, Chad Kennedy, Steve Maslin, Hal Moggridge, Professor Graham Rook, Alan Turner and Ben Ward for allowing me to use extracts of their work.

Jeremy Barrell, the Centre for the Protection of National Infrastructure, Considerate Constructors, Jackie Cross, David Jarvis Associates, Hani Hatami at Humanscale, Ilex Paysage + Urbanisme, the International Living Future Institute, the staff at the Landscape Institute archive and library, the Phipps Center for Sustainable Landscapes, Talley Associates, Lucy Wiltshire and colleagues at Morgan Sindall Construction & Infrastructure Ltd and what3words for so generously providing photographs.

The Get It Done Group – Barrie Atkin, Chip Capelli, Gwyneth and Winifred de Cavia, Marlyn Keating, Bridget Keller, Catherine King, Joyce Kristjansson, Beth Perry and Stever Robbins for being such wonderful cheerleaders and proving that collaboration doesn't need to be face to face to work.

Charmian Beedie, Richard Blackburn, Clare Holloway, Susannah Lear, Sian Parkhouse, Liz Webster and everyone at RIBA Publishing for expertly and reassuringly getting this book from idea to publication.

My wonderful clients who allow me the privilege of working on their landscapes

And my family – no, I won't be selling the film rights.

Given my background this book is written from a UK perspective but I hope that landscape architects in every country will find relevance to their work. If your country has solved any of the issues I raise, or in your work you approach things from an entirely different perspective, let me know. I want this book to inspire discussion, for us to question the way we work and to ultimately create even better landscapes.

For some of the points I raise I don't offer solutions – I've included them as I want to highlight issues and then allow you to draw your own conclusions.

My hope is that reading this book will allow you time to think about your work – even if you disagree with me on every point you will have taken time to consider your work, why you do it and hopefully how you can become an even better landscape architect. If I achieve just that, I will be happy.

Claire Thirlwall, Oxfordshire, 2019

Landscape architects work on the spaces between buildings – it is likely that as many people engage with the work of landscape architects as that of architects, yet the profession receives little coverage in the media or the construction press, with a disparity in the values assigned to architecture and landscape.

Whether due to a lack of understanding of the skills we offer, or a lack of value placed on the results we can achieve, some clients see the work of their landscape architect as a low priority. This book aims to counter this view, using case studies to illustrate the diversity and value of our work, such as how landscape design can counter terrorist attacks, help sites adapt to climate change or move to a more sustainable, low-carbon style of maintenance. It also looks at innovations that can help improve the quality of the information we use to make design decisions, such as cloud-point scans and open data.

Rather than focus on the finished landscape scheme, with glossy photos that belie the work taken to achieve the result, the book looks at the detail of the work required to take a landscape project from idea to site. Using the RIBA Plan of Work as a framework, each work stage is explored from the perspective of the landscape architect, the client, the project team, the project and wider society.

From Idea to Site explores working practice – can new technologies help us work more effectively, and how can we show clients the full range of our skills? It is a guide rather than a reference book: it highlights new ways of dealing with challenges in the profession, hopefully helping landscape architects reach new conclusions about their working practice, in turn improving the quality of the design and management of external environments.

The book uses lectures, podcasts, articles and books as source material, as well as personal conversations and project experience. The events I have attended have been the greatest source of inspiration, either from the speakers or in the post-event conversations, demonstrating the value of taking time away from project work to look at new ideas and make new discoveries.

For simplicity, the structure and terminology is based around projects working in the private sector with multi-disciplinary teams, but the process is equally applicable to all landscape architecture projects, including the public and charitable sectors.

I hope that you disagree with some of the points I make – I'd like this book to prompt discussion and to encourage us to question the way we work. There are always improvements to be made and we should never assume that we have all the answers.

Chapter One
DEFINE

INTRODUCTION

Before a project begins, before any of the formal works stages are considered or even before the first conversation with the potential client, decisions are made that determine success. Some are outside the control of the landscape architect, such as standards set by legislation. However, many of the standards we work to are up to us – our values, our skills and our aims for the project. Identifying and defining these standards can help us decide the projects we work on, the clients we work with and the values we wish to uphold.

Some decisions or standards may be implicit, only discussed if we are asked to work on a controversial project. We might be certain that we would never work on a project directly linked to an extreme political party. But how far would we take that view – would we work with their supporters? Would we choose suppliers who donate to campaign funds? Would we have an account with a bank that finances them? Companies might not be open about all their business practices, so we need to decide what our own thresholds are, and how much research we can realistically undertake. Credit checks and other due diligence checks are part of taking on a new client, but should we check their conduct and allegiances?

This chapter explores some of the standards and criteria we might work to and the guidance available to help us make the best decisions from the outset of a project.

Fig 1.0 (Pages XII and 1) View from Beacon Hill Country Park towards Ratcliffe-on-Sour Power Station, 2017, Woodhouse Eaves, Leicestershire, UK

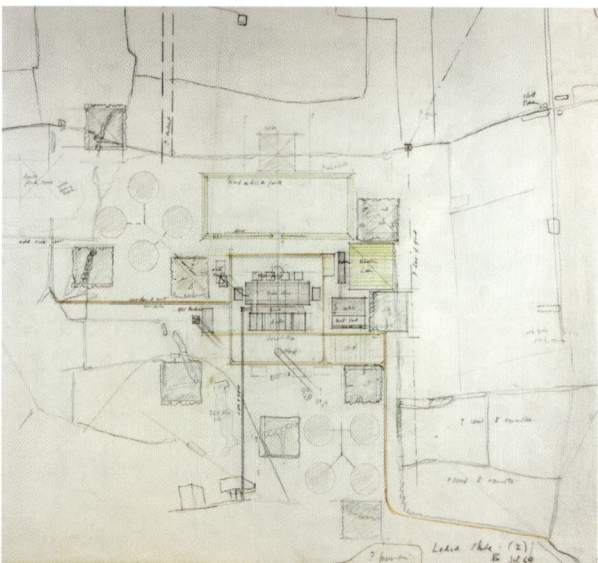

Fig 1.1 Didcot Power Station the evening before demolition of the three southern towers on 26 July 2014 – landscape architects work on large-scale projects that have a profound impact on the landscape; Frederick Gibberd for Central Electricity Generating Board 1964

Fig 1.2 Landscape study for Didcot Power Station (Didcot A); sketch plan by Frederick Gibberd 1964, Sutton Courtenay, Oxfordshire

PRIOR TO STAGE 0
STRATEGIC DEFINITION

– Before a new project is instigated

STAGE 6 – HANDOVER AND CLOSE
AND STAGE 7 – IN USE

– Use client feedback and lessons learnt from previous projects to review standards and refine processes for future projects

Table 1.1 The elements of this workstage[1]

THE LANDSCAPE ARCHITECT

What sort of
landscape architect am I?

How often have you stopped and thought about what sort of landscape architect you are? Maybe you've questioned your principles when faced with a moral dilemma, such as a controversial project or a challenging client, but how much time have you spent considering your approach to your work?

Why does it matter?

Knowing our own standards and defining what sort of landscape architect we are, and want to be, matters. It means we know what projects we are prepared to take on but also how we want to work and if there are areas in which we'd like to specialise. We might be able to work for short spells on projects that are at odds with

our beliefs, standards and ethics. In the longer term working on projects that you don't support is at the very least unfulfilling, meaning we lack enthusiasm, motivation or creativity, or don't treat the project as a priority. At worst it could be a risk to our reputation, livelihood or even our health.[2]

It may sound like an indulgence, far removed from the immediate issues that we face with our projects. However, knowing your own parameters – what you will and won't accept – is an important step. At a most basic level it means that you won't work on projects that are odds with your own beliefs, a situation that rarely results in a successful project. As we progress through our careers the type of landscape architect we are is likely to evolve, and if our circumstances change we may have to review our approach entirely. Understanding the type of landscape architect we are, and that we aspire to be, is central to all elements of our work, and should be considered at the outset of any project.

Whilst writing this book I have asked other landscape architects if there are projects they would refuse to work on, or clients they would never consider. Some said they would work on any project, but when pushed it was clear they had limits that they had never considered. Others had specific activities, such as fracking or new housing, that they would not want to work on. Discussing personal views and how employers accommodate these has been fascinating.

For sole practitioners the situation is perhaps simpler, as you only have your own opinion to consider, but with greater potential consequence. In a large practice the impact of declining the occasional project on ethical grounds is likely to be minimal. However, for

Fig 1.3 Great Western Park housing development – some professionals choose not to work on new housing schemes on greenfield sites, various developers, 2019, Didcot, Oxfordshire

a smaller practice where each job is a higher proportion of income, turning down much-needed work can be a difficult decision.

It is a complex issue, and even a project that we are happy with in principle may have elements that we are uncomfortable with. Given the nature of our work and the range of projects we can be involved in it would be a mistake to set hard and fast rules, as each situation needs to be judged in context. Our codes of professional conduct might give us some guidance, but these still leave a great deal of latitude for a landscape architect to define their own approach.

Landscape architecture has so many facets that there should be a rewarding option for most of those who take up the profession. Working in a field that you enjoy means you are likely to come across as authentic, able to support your design decisions to your client and outside parties. Running a public consultation event or giving evidence at a public enquiry for a project that you don't fully agree with is an uncomfortable experience. A project may be within the scope of our code of conduct and have no ethical concerns, but still not sit comfortably with our ideals. As a professionals we don't always deliver the project in the way that we would have chosen, as our task is to carry out the client's wishes, which may not be the same as our own. However, working on something that is very far removed from our ideals is unlikely to lead to a successful project outcome.

Defining our own standards

How can we develop our own personal set of standards, and what guidance is available if we want to explore this topic?

Universal standards

Each society has a set of standards that define what is acceptable conduct – serious standards are usually enforced by law, less serious ones determined by social norms. Standards such as legislation define the lowest acceptable standard – they define the least that is expected of us. Standards defined by social norms apply to business as well as personal conduct and can differ from country to country, or even within different sectors of the same business, depending on context – compare the formality of a board meeting with an informal site visit.

An interesting standard to explore is the United Nations Global Compact, created as part of the 2030

Agenda for Sustainable Development.[3] The Compact, supported by over 9,000 companies in 161 countries, has 10 principles covering human rights, labour, environment and anti-corruption.

Business standards

We need to comply with the legal requirements of running a business or organisation but there are other factors that need to be considered, all of which can impact on staff performance, financial viability, and the subsequent success of any project. These include payment terms, profit margin, financial risk, pay scales and employment terms, such as zero-hours contracts or unpaid overtime or internships. Less than desirable standards are rife in construction, so it pays to be diligent.

Construction sector standards

The construction sector has a number of recognised standards that can be applied to projects, including those worked on by landscape architects. Most standards focus on the construction of buildings, with the landscape element of the project playing a minimal part in the criteria, but there are a few standards designed for landscape schemes, or that include a significant emphasis on our work. These include:

Living Building Challenge – a standard for buildings, infrastructure or landscape projects. Set up in 2006 the standard includes ethical, community and environmental elements. The Challenge, which sets ambitious aims, including the requirement to be net positive in relation to water, energy and ecological regeneration, is described as 'a philosophy, certification and advocacy tool for projects to move beyond merely being less bad and to become truly regenerative'. The standard is not intended as a mass-market certification – instead it aims to encourage innovation and advocacy to set an example to the sector and show what is possible. The standard can be applied to landscape and infrastructure projects as well as buildings.[4]

Sustainable SITES Initiative (SITES) – provides certification for projects with or without buildings, including national parks, streetscapes and homes. The emphasis is on sustainable landscapes. Originally set up in collaboration with the American Society of Landscape Architects (ASLA) the scheme is now run by Green Building Certification Inc., which also manages the LEED and WELL certification standards. Like LEED the standard has Certified, Silver, Gold and Platinum levels of certification.[5]

Personal standards

Our personal standards and beliefs set the tone of how we will work on a project from the outset. It is useful to reflect on the deep-seated attitudes that impact on our work. Points to consider might include:

Working hours – Are we willing to work long hours to complete the project? This decision may be outside our direct control, and we may be prepared to extend our working hours in exceptional circumstances but are we (and should we) be prepared to work longer than our contracted hours to deliver our work? Research by the Finnish Institute of Occupational Health shows that long working hours can lead to increased risk of cardiovascular disease, diabetes and increased risk of heavier alcohol use.[6] Lack of sleep can impair our ability to read people's emotions and make decisions.[7]

It also puts us at risk of accidents – in the UK driver fatigue was thought to be a contributory factor in 20% of all road accidents and up to 25% of all fatal and serious accidents.[8] Tiredness is thought to have been a contributing factor in the Challenger Space Shuttle disaster, the Chernobyl and Three Mile Island nuclear incidents and the Exxon Valdez oil spill.[9] Evidence suggests that longer working hours rarely result in greater productivity and can lead to serious harm, particularly over a sustained period, but in many sectors long hours are seen as a commitment to our job and a mark of importance. Considering the personal cost of projects should be part of our review process – is the potential harm a valid price to pay?

Conduct – How do we expect our colleagues and our clients to behave? Are staff valued or are unreasonable demands made? Is bullying commonplace, or harassment accepted? Legislation may be in place that aims to tackle workplace harassment but it doesn't address workplace cultures that support bullying or prejudicial behaviour.

Gender discrimination – Are we biased in how we recruit and promote staff? Again many countries have legislation to prevent gender discrimination but the issue persists. A 2017 survey by the UK Landscape Institute showed a gender imbalance in senior roles with 20.4% of male respondents earning over £50,000 per year compared to 9.4% of female respondents.[10] The reasons behind this disparity may be complex but we need to consider the part discrimination plays in restricting the senior management opportunities for female landscape architects.

Workplace culture – Landscape architecture practices might not follow the tech industry trend for quirky workplaces but as creative professionals the culture of our workplace can impact on our creativity. The main factors in workplace creativity aren't beanbags or table football – these are more about making extreme working hours bearable by blurring the line between work and leisure.[11] More important factors for our work are the level of autonomy given to solving problems, setting realistic and genuine deadlines, and accepting that unsuccessful efforts are part of the creative process. As well as our personal beliefs, how we perform at work is determined by our previous experiences, and the judgements we have made based on those experiences. Addressing and reviewing these personal judgements is outside the scope of this book, but some useful publications are listed in the Further Reading section at the end of this book.

Fig 1.4 (Page 6) Sustainable SITES Initiative accredited landscape scheme at the Perot Museum of Nature and Science – the project was built on a brownfield site and the irrigation demands are met by recapturing air conditioning condensation; landscape scheme by Talley Associates 2013, Dallas, Texas

DEMOLISHING GENDER INEQUALITY IN THE CONSTRUCTION SECTOR

In December 2016 the Australian Sex Discrimination Commissioner and University of New South Wales released the findings of their study into gender inequality in construction. The report, the result of 2 years' work including 61 interviews and work-shadowing 44 construction professionals, showed that both men and women fare badly because of existing working practice. The construction industry is the most male-dominated sector in Australia, with women representing only 16% of the workforce and just 14% of professional and managerial roles. The report found that initial enthusiasm for working in construction professions decreases with exposure to the workplace, resulting in women leaving the sector 39% faster than men. Findings included the following points:

- Recruitment – Male sponsorship: Recruitment onto projects routinely operates through a practice of male sponsorship and 'picking your team'. This undermines diversity of talent and limits women's access and opportunities in the industry.
- Retention – Exclusion: The exclusionary nature of the construction industry operates to remind women – subtlety and overtly – of their gender and difference; these reminders frustrate and exhaust women over time. There is a tolerance for sexism in construction – sexist comments, sexist graffiti, asking women to do administrative work, and other practices that make women feel they are intruding in a male-dominated space.
- Progression – Undermining women's capabilities: Men's capabilities as a construction professional are assumed; women's capabilities are frequently questioned, singled out or discussed. Women need to better, not equal men. Actions to address gender equality are viewed by men as providing women with an unfair advantage.

The report recommended several steps to help reduce inequality and to improve working practices in our sector:

Recruitment

- Make company and project recruitment processes and criteria more transparent.
- Review the values that underpin 'cultural fit' to determine if they are gendered and exclusionary.
- Initiate recruitment drives specific to women not from the traditional pipeline and provide these recruits with construction training.

Retention

- For women, it is important to see other women in senior ranks and be placed with other women professionals on site.
- Stop rewarding and promoting excessive hours and 'shaming' those who don't comply with excessive hours.
- Introduce job sharing. Standardise work hours. Remove Saturday work. Monitor fatigue. Talk about it. Enforce it.
- Demonstrate 'no tolerance' to sexism – sexist drawings, wording, behaviour – in the workplace (including the site).
- Endorse parental leave practices 'on the ground'. Introduce the option for staged return to work for parents.
- Set up projects with gender diversity in mind. Plan for flexibility, wellbeing and parental leave.

Progression

- Make promotion processes and criteria more transparent.
- Change the narrative. Recognise, recruit and celebrate agile and diverse career pathways and career breaks.
- Establish a formal sponsorship programme for women in low to middle management.

THE FIRST FEMALE LANDSCAPE ARCHITECT

Fanny Rollo Wilkinson (1855–1951) is credited as the first woman to train as a landscape architect, graduating from the Crystal Palace School of Landscape Gardening and Practical Horticulture in 1883. Fanny designed over 75 public sites across London in her role as honorary landscape gardener to the Metropolitan Public Gardens, Boulevard and Playground Association as well as her work with the Kyrle

Society, an organisation founded to bring beauty to the lives of the poor.

Her projects included Vauxhall Park and Myatt's Field in Hackney, London. In 1904 she became the first female principal of Swanley Horticultural College, the college from which famous landscape architects Sylvia Crowe and Brenda Colvin, the first female President of the UK Landscape Institute, later graduated.

Fanny was an active member of the women's suffrage movement:

'I certainly do not let myself be underpaid as many women do. There are people who write to me because I am a woman, and think I will ask less than a man. That I will never do. I know my profession and charge accordingly, as all women should do.'

MENTAL HEALTH IN CONSTRUCTION

The mental health statistics for the UK, and for the construction sector in particular, are startling. Statistics show that in the UK:

- Men aged between 45 and 49 have the highest suicide rate.
- Men are three times more likely to take their own lives than women.
  Sector-specific details reveal that:
- The risk of suicide among low-skilled male labourers, particularly those working in construction roles, was almost three times higher than the male national average.
- For males working in skilled trades, the highest risk was among building finishing trades; particularly, plasterers, painters and decorators

had more than double the risk of suicide than the male national average.[12] [13]

There are some positive signs, with male suicide rates in the UK at their lowest level for 30 years. However, a 2018 report by the Work Foundation cites diverse factors such as working in high-risk environments, low pay and job insecurity, working away from home for extended periods, the 'macho' image of construction, long commutes not allowing for adequate sleep or exercise, higher risk of engaging in risky health behaviours such as drug and alcohol abuse and a reluctance to engage with health services as potential contributing factors. However, the report concludes that the reasons for this increased suicide risk are not fully understood.[14]

The report highlights several initiatives including the UK-based Mates in Mind. Established in 2016 and prompted by research, the campaigning organisation works to provide information for employers on support and guidance on mental health, mental illness and mental wellbeing. The charity works in partnership with existing organisations such as Mind, Samaritans and Mental Health First Aid England.

The main aims of the organisation are to:
- raise awareness and understanding of mental health and mental ill-health
- help people to understand how, when and where to get support
- break the silence and stigma through promoting cultures of positive wellbeing throughout the industry.

They run Mental Health First Aid training, the www.matesinmind.org website with links and resources, and social media campaigns including #GetConstructionTalking.

The statistics throw up a wider issue: that working practices within our sector have serious and potentially life-threatening implications. Low profit margins and tight timescales don't just mean that employees and sub-contractors are financially vulnerable, they also place a strain on the mental health of those with the least say in their terms and conditions.

If we are working with a group of people whose idea of good behaviour is at odds with our own, making our working environment a challenging place to be, we are unlikely to deliver our best work.

THE INTERNATIONAL FEDERATION OF LANDSCAPE ARCHITECTS

Founded in Cambridge in 1948 by UK landscape architect Sir Geoffrey Jellicoe the not-for-profit International Federation of Landscape Architects (IFLA) represents 76 national associations, officially representing 25,000 landscape architects across the world.

The Federation was created in the same era as the United Nations and the Universal Declaration of Human Rights and was part of a post-war movement that strived for a better future. Landscape architects were considered central to the rebuilding of war-damaged landscapes. The Federation works to promote the work of landscape architects as well as providing information for leaders on the role of landscape to meet challenges including water and food security, climate change, migration, housing, conflict and the depletion of resources.[15]

IFLA Europe has a Code of Ethics and Professional Conduct that covers personal attitudes, professional competencies, professional relationships and the environment and is included in full in the Appendix.[16]

Standards in landscape architecture

In our work as landscape architects we have responsibilities to a number of parties, in addition to our responsibility to our clients. We need to consider diverse issues such as the impact on the environment and the needs of those with an interest in our work. If implemented well our work could be in place for many generations, so it is important that we aren't pressured into actions that jeopardise the future potential for a site.

International comparison

Standards for landscape architects may not translate from region to region or country to country, and we should celebrate the diversity of approach, but there are common elements that all landscape architects work to.

CODE OF CONDUCT – OVERVIEW

As professionals we can assume we know what standards we work to. Our professional body may require us to sign up to a code of conduct as a condition of membership. These codes, such as the UK Code of Standards of Conduct and Practice for Landscape Professionals, set out specific requirements such as financial conduct and professionalism.[17]

UK code of conduct

The UK Landscape Institute Code of Conduct 'expects members who are carrying out professional work to have regard to the interests of those who may be reasonably expected to use or enjoy the products of their work'. This could include clients, employees, investors, suppliers and the public, as well as ourselves. We need to consider users of sites, present and future. With so many potential users it can be difficult to decide who is a priority, and there can be conflicts between different interests.

Codes of conduct don't ensure higher standards of behaviour are always complied with, but they do publicly set out the minimum standard that should be expected by a client from a professional. They also help create a fairer environment when bidding for work – if all landscape architects are bound by the same standards, such as undertaking the same level of continuing professional development (CPD) and not allowing conflicts of interests, there should be parity.

International

The content and scope of each code of conduct varies significantly from country to country, in part depending on how recently they were written but also on the professional priorities in that country. For example:

- The American Society of Landscape Architects has two sets of standards for members to adhere to – a Code of Professional Ethics covering working practice and a separate Code of Environmental Ethics.[18][19]
- The Australian Institute of Landscape Architects prohibits any payment of commission or other benefits outside of a professional fee.[20]
- Sveriges Arkitekter, the professional body for architects, interior architects, landscape architects and spatial planners in Sweden, instructs members to 'be loyal to work colleagues, contribute to promoting an open and creative climate in the

workplace and otherwise make a commitment to a democratic approach'.[21]
- The Institiúd Tírdhreacha na hÉireann, the Irish Landscape Institute, includes a clause stating that 'Members shall strive to accomplish the objectives of their work with the most efficient consumption of natural and manmade resources, including the maximum achievable reductions in energy usage, waste and pollution.'[22]

Codes of conduct and legislation may be a basis for professional standards but they are still only the absolute minimum required of us, leaving scope for our own interpretation of what professional values mean. Minimising company tax bills, paying low wages or using zero-hours contracts are all legal but are seen by some as unethical. Working as a professional requires us to define our own set of standards, standards that will change and adjust depending on context, and that are personal to us.

ETHICS IN LANDSCAPE ARCHITECTURE

Some of the issues we can come across in our work are shown below:

Modern slavery/ forced labour	Bribery and corruption	Unfair tender practices	Payment terms	Toxic materials	Funding sources
Accessibility and exclusion	Public/private space	Diversity	Working hours	Profit margins	Habitat species loss
Child labour	Pay scales	Gated communities	Misleading marketing	Human rights	Offensive behaviour in the workplace
Discrimination	Environmental responsibility	Development in protected landscapes	Membership of professional bodies	Nurture young talent v costs	Fracking

Table 1.2 Issues to consider

THE CLIENT

The stage we are exploring doesn't include a specific client, but we can review the type of client we might work with, which is especially useful to consider when targeting new areas of work. We might only work with clients with fair payment terms, or who come to us by personal recommendation. We might also set criteria based on a client's standards and values, perhaps declining clients whose poor reputation might affect our own, or who openly support causes we don't agree with.

What is a good client?

Le Corbusier said: 'To accomplish good design you need a good client.' Aside from the fundamental factors, such as signing a contract and paying invoices on time, there are other traits typical of a good client.

Making any generalisation is never a good idea, but typically a good landscape architecture client:
- values the skills of the professionals they appoint, considering each profession to have an equal role and respecting expertise
- recognises that paying for advice and expertise may increase design stages costs but reduces the risk of changes or delays during construction
- understands that landscapes are dynamic – they change and develop over time, and they may look their best hundreds of years after completion
- appreciates that landscape architecture is subject to natural processes that can't always be mitigated, such as slow plant growth, low rainfall or plant disease
- contributes to the design process and isn't afraid to ask for the rationale behind design decisions

- communicates effectively, raising issues as they arise and giving clear and timely decisions
- sets out and implements a clear line of instruction and reporting, with clear roles for those linked to the project
- agrees contract terms, either with a formal contract such as the Landscape Consultant's Appointment, or with bespoke terms tailored to that project. Not having an agreed method of dispute resolution, no-fault termination or even proof of appointment is a foolhardy way to run a project
- provides all information relevant to the site. In the UK the client is legally required to divulge all information that relates to Health and Safety, such as site contamination.[23]
- pays on time – a profitable business can fail due to poor cashflow. Appropriate payment terms and a process for monitoring overdue payments can help, but a client who always pays on time is the best insurance against poor cashflow. The March 2016 Euler Hermes Quarterly Overdue Payments Report showed that construction companies registered more payment delays than any other sector, with a 26% year-on-year rise in 2015.[24]

Some clients, especially those with limited experience in the role, may need our help to become a good client.

CARILLION PAYMENT TERMS
IS 120 DAYS A FAIR STANDARD?

When the UK construction and services company Carillion collapsed on 15 January 2018, thousands of jobs were put at risk and hundreds of millions of pounds of public contracts left unfinished.[25] In spring 2017 the company had been signed off as a going concern by professional services company KPMG but at the point of collapse the company held just £29 million in cash and at least £5 billion of debt. This imbalance had been supported by extended payment terms for their supply chain of 30,000 companies – in March 2013 the firm had implemented an early payment facility that required suppliers to accept 120 day payment terms.[26] With some public-sector projects making a substantial part-payment up front, Carillion could use this new cash to pay existing debtors, but the process relied on continuing to win new contracts.

Payment terms set out in UK government contracts are 30 days, but the complex early payment facility set up by Carillion only allowed payment in under 120 days if the supplier accepted a sliding scale of charges depending on how early they wished to be paid.[27] Carillion was a signatory to the Prompt Payment Code which requires signatories to undertake

to pay suppliers within a maximum number of days, and to avoid any practices that adversely affect the supply chain.[28]

The scheme was a form of supply chain financing, with banks covering the early payment for a fee. Though entirely legal, and a practice endorsed by the government at the time, the practice placed the risk with the supplier, demonstrating the asymmetry of power that an unequal supply chain relationship can support. The payment was in reality an unsecured loan between the bank and the supplier, so when Carillion collapsed the banks went to the suppliers to recover the money.[29]

According to the report by the UK Government select committee tasked with investigating the collapse:

'Carillion relied on its suppliers to provide materials, services and support across its contracts, but treated them with contempt. Late payments, the routine quibbling of invoices, and extended delays across reporting periods were company policy. Carillion was a signatory of the Government's Prompt Payment Code, but its standard payment terms were an extraordinary 120 days. Suppliers could be paid in 45 days, but had to take a cut for the privilege. This arrangement opened a

line of credit for Carillion, which it used systematically to shore up its fragile balance sheet, without a care for the balance sheets of its suppliers.'[30]

As the company faced collapse Carillion considered a proposal to extend payment terms to 126 days as an untapped 'cash generative opportunity'.[31]

In the months since the collapse of the company further issues have become known. A report by Oxfordshire County Council, a UK local authority whose remit includes schools and roads, revealed missing health and safety manuals and building control certification as well as unresolved planning conditions and unsatisfactory fire strategies.[32] As a *Construction News* article 10 months after the collapse commented:

'The audit poses the question: would these issues ever have been discovered if the council hadn't been forced to look into the practices of one of its major service providers?

How many other corners were cut that are only now being discovered, with those responsible hoping the problems would never see the light of day?

It should not need things to go tragically wrong in construction for these problems to be unearthed.'[33]

THE PROJECT

We probably have an instinctive response to projects, knowing which potential projects would be too small to allow any profit, or too time-consuming to fit with our existing workload. As with other decisions the extremes are easy to assess – it is the grey area between the extremes that is harder to make judgements about.

Working through some of the project parameters and deciding a set of criteria against which to select projects can help rule out unsuitable projects and decide which areas of landscape architecture you want to focus on.

Some criteria to consider include:

Value – What is the lowest value scheme you will work on? Or the highest? High-value schemes are not always complex and high values shouldn't be a deterrent even for smaller practices. However, higher contract values will require higher insurance. Understanding the costs to set up a new project can help calculate a maximum and minimum contract value. There can always be exceptions, such as a small project for a community group, but the decision to take on pro bono work must be based on a full understanding of the potential implications.

Location – How much travel time is realistic for a project? A project might be a great opportunity but if it involves excessive travel it could be hard to manage and impact adversely on your personal life. What are acceptable demands on staff? Will the location make meeting start times unrealistic? Is the site safe for staff to visit? Would you work in areas considered dangerous by your government, perhaps to carry out a humanitarian project? Work such as landscape character or visual impact assessment can sometimes require large amounts of driving, especially if the landscape has few roads, and the nature of our work often means that using public transport is not an option. Location and time of year can mean a project isn't viable – limited hours of daylight and a remote location could mean that a job requires the extra cost of an overnight stay, when in the summer it could be completed in one day.

Expertise – What are our areas of expertise? Do we have a specialism that we want to focus on, such as habitat recreation or housing, or are there gaps in our expertise we need to address? Are there types of project we would always refer to another landscape architect?

Reputation – How will this project influence our professional reputation? Might a certain type of project deter some potential future clients due to the kind of project, or, conversely, does it offer the opportunity to enhance our reputation, raise our profile or improve our skills in a particular area?

Workstage – What is the latest point we are prepared to join a project? Being called in to discharge planning conditions allows limited scope to influence a project, but it may be a chance to work with a new client in the hope that they appoint at an earlier stage once they understand the value of our work. I don't feel that landscape architects can be appointed too early – being involved at a strategic stage means that visual impact, structure planting, access layouts and any supporting infrastructure can all be considered before irreversible decisions have been made. The stages

HOPE VALLEY CEMENT WORKS, DERBYSHIRE, UK

Deciding which projects to take on can create moral dilemmas, especially when the site is in a protected landscape such as a national park. In the 1940s landscape architect Sir Geoffrey Jellicoe worked on the Hope Valley Cement Works, a major industrial site in the heart of the Peak District National Park and still in operation today. Working on such a visually intrusive scheme in such a sensitive landscape may seem an unusual decision for a practice to take, but mineral resources are often found in protected landscapes, meaning that the landscape impact

has to be balanced against the national need for that resource – post-war reconstruction created huge demands for building materials.

The works, then known as Earle's Cement Works, were discussed in 1949 in the UK Parliament. Consent had been given for a large extension to the site, and the wider issue of mineral extraction and the role of the proposed national parks in the control of this process was discussed. Despite the objections to the visual and physical impacts of the Works Jellicoe's role was praised by Hugh Molson MP, then Member of Parliament for the High

Peak constituency. It is interesting that he is referred to as a landscape artist:

'The first example to which I want to refer is the permission which he granted after a public inquiry for a great extension of the existing cement works in the Hope Valley. I appreciate fully what can be said in favour of what is there being done. I know that Earle's Cement Works have not been indifferent to considerations of amenity. Quite a long time ago they engaged Mr. G. A. Jellicoe, one of the most distinguished of our landscape artists, to see what could be done to make reparation and to cover up the scars which have been inflicted upon the Hope Valley, and the harm that has already been done to the amenities by their industrial development. I was very glad to open an exhibition in the village of Hope showing in a small plan the scheme which had been prepared by Mr. Jellicoe and which Earle's Cement Works had undertaken to put into effect.'[34]

New building
Waste disposed or
to be disposed
Quarries
New planting
Property line
Critical Silhouettes
Hill from waste grasses
and tree planting
Terrace from waste
Mature Trees

Fig 1.5 Landscape as existing plan – extract from report 'Hope Works Derbyshire – A Progress Report on a Landscape Plan',1979; produced by Geoffrey Jellicoe for Blue Circle Industries 1993, Castleton, Peak District National Park, Derbyshire

and potential influence of our work are covered in Chapters 3 and 4.

Drawing up guidance on what types of project and ways of working are acceptable, or even just taking time to consider the projects we would or wouldn't work on, helps us maintain a consistent approach to our work, and prevents us taking on work at odds with our values. Understanding that we might have to weigh up complex and seemingly contradictory issues, and that our response may change depending on the context, is a core skill of a professional landscape architect.

THE PROJECT TEAM

For many projects a bespoke team is created, and the composition of that team may be outside our control. However, we may be able to advise our client as to the other professions needed to create an appropriate project team. Understanding the scope of each profession's work, including what is outside the scope of our own work, is important. The UK Landscape Institute Code of Conduct prohibits landscape architects from misrepresenting their skills, and professional indemnity insurance might not cover work outside our normal area of expertise.

The tendency can be to work with the same team, especially if that team has been successful on previous projects. However, working with an established group can lead to complacency and discourage innovation. It also makes it difficult for new practices or younger staff to gain experience if new team members are excluded.

A project team made up of people with similar values, such as matching expectations of acceptable working hours or response times to emails, can reduce conflict within a project team. However, research suggests that a team with no conflict can lead to less creative outcomes and mean that issues aren't dealt with.

Working with colleagues who share our wider values, such as wanting to minimise the impact on the environment or maximise the social value of our work, can help make our work rewarding.

Who is in the project team?

The diversity of our practice means we can work with a wide range of other professionals. We often work with ecologists, arboriculturists and archaeologists to assess the site before the design stages begin, collaborate to design mitigation and to monitor the site during construction.

During the design process we might work with a whole range of professions, ranging from geo-technical specialists to the mechanical and engineering (M and E) consultants. It can be hard to explain to a client why we need to work with the entire project team, especially if they see our work as a peripheral item and not integral to the whole project. However we do need to understand how all the planned activities, during construction and in use, might interact with our landscape scheme. Locations of underground services, placement of vents, pipework or ducts, overhead lines or sight lines for safety and security are all factors we need to accommodate in our design. We also need to know any parameters we need to work to – examples include final tree height restrictions close to a wind turbine or the risks from tree roots to pipelines, structures or surfaces.

Within project teams the tendency can be to only provide the landscape architect with information on a limited range of topics, based on the misperception that our work has a limited scope. This is misguided, as we need to understand the entire project, including planned maintenance, to create a successful design.

Disciplines and sectors we might work with include: access; acoustics; air quality; archaeology; architecture; BIM; civil engineering, including structural, geotechnical, transport and environmental; contamination; cost consultancy; ecology; environment; facilities management; fire engineering; health and safety; interior design; lighting; planning; soils; security; surveying, transport and traffic.

WIDER SOCIETY

As the UK Landscape Institute Code of Conduct points out, as well as a responsibility to our clients, we also have a responsibility to those who might use or enjoy our work. As mentioned previously, this includes future users.

An interesting reference for our work is the Well-being of Future Generations (Wales) Act 2015 brought in by the Welsh government. The Act, the first piece of legislation in the world to enshrine sustainable

Fig 1.6 Incredible Edible planting bed in the snow – small-scale community project to grow food in an urban setting; managed by Sustainable Didcot 2018, Didcot, Oxfordshire

Fig 1.7 Street trees can reduce air pollution – urban trees along Strandvägen, Stockholm, 2006

development in law, forces the public bodies listed in the Act to think about the long-term impact of their work. The legislation covers topics including climate change, poverty and health inequality and places an emphasis on preventing problems occurring or getting worse.[37]

The impact of our work is potentially wide ranging, requiring us to consider topics such as:

— **climate change adaption** – sequestering carbon within our projects, selecting plants that can tolerate changing conditions such as lower rainfall or higher summer temperatures

— **food production** – maximising the use of agricultural land, and finding ways to include food production that fits with the urban fabric whilst being practical and productive

— **air quality** – vegetation, especially trees, can help remove pollution from the atmosphere, filtering out the smallest particulates that are a major risk to health. Species need to be chosen carefully to maximise the effect, with dense canopies and small or hairy leaves providing the best results

— **terrorism and security** – incorporating structures to prevent vehicle attacks and designing layouts to allow natural surveillance and discourage anti-social behaviour.

Our work might not be widely recognised by the public but many of our projects impact on their lives and they should be central to our decision making.

Choosing to ignore an issue is a decision in itself – not having a view does not make that issue go away or reduce the risk of its impact on your work. Avoiding an

FOREST STEWARDSHIP COUNCIL ®

In the early 1990s a group of timber users, traders and representatives of environmental and human rights organisations met in California. Prompted by concerns over deforestation, environmental degradation and social exclusion the group identified the need for a credible system for managing and sourcing responsibly produced timber products.

Two years later, at the 1992 UN Conference on Environment and Development – the Earth Summit – Agenda 21 was drawn up along with the non-legally binding Forest Principles. This led to the creation of an assembly that in turn developed a worldwide certification system. The Forest Stewardship Council® (FSC®) become a legal entity in February 1994.

The impact of the standards created and managed by the FSC are wide ranging, covering all wood products. In 2007 *Harry Potter and the Deathly Hallows* was printed on FSC certified paper – the single largest FSC paper order, valued at $20 million.[35]

The progress has been impressive, but there is still work to do. Whilst the requirement for FSC certified products is common in the UK construction sector, in 2017 only 17% of the world's production forests are FSC certified.[36]

issue can lead to a damaged reputation, or unacceptable behaviour carried out in your name.

We may make different decisions based on context – during a recession we might accept payment terms that we'd decline in better times – but professionalism is about defining standards that we can justify to ourselves and our clients, and that we will never break regardless of context. Standards can change over time and previously unacceptable behaviour becomes unacceptable.

Our motivations for working to defined standards may be altruistic, putting the needs of others before our own. Our motivations can also be more pragmatic – staying within the law, paying our bills or maintaining a good reputation. We can't just consider the interests of the people and organisations who commission our work – the work of landscape architects usually has a long-standing impact and our work may far outlive any surrounding buildings or infrastructure.

We can decide who we will work with and the type of work we will take on, the way we will work and the compromises we will accept to ensure a project is a success. Even if we have no formal criteria, we will have an idea of what project values, timescales or locations are realistic. We may make different decisions based on context – during a recession we might accept payment terms that we'd decline in better times, but professionalism is about defining standards that we can justify to ourselves and our clients, and that we will never break regardless of context.

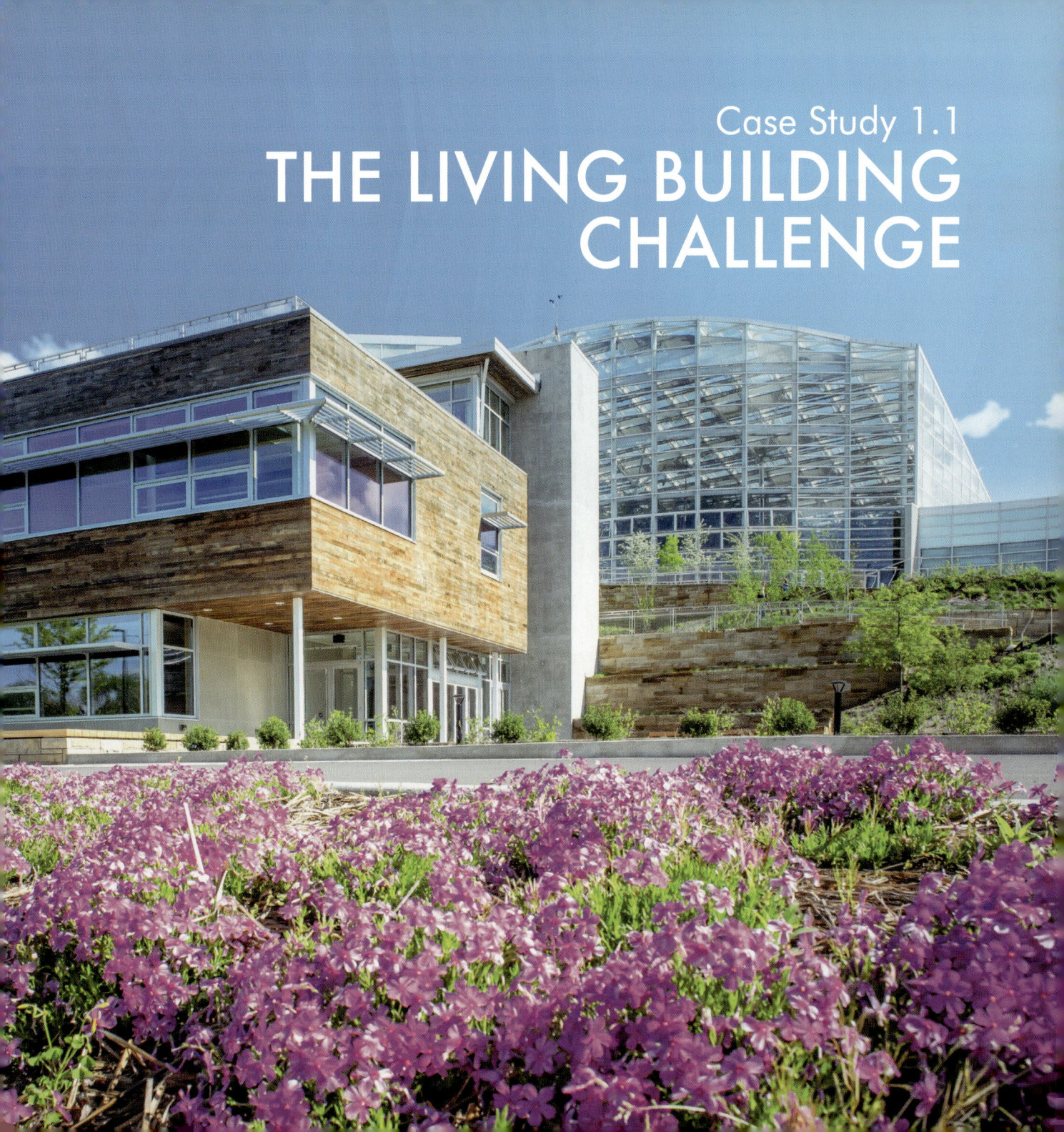

Fig 1.1.0 (Page 21) A rain garden blooms in front of the entrance to Phipps Center for Sustainable Landscapes; landscape design by Andropogon Associates for Phipps Conservatory and Botanical Gardens, 2012, Pittsburgh

Fig 1.1.1 (Page 22) Native planting at Phipps Center for Sustainable Landscapes. The project achieved Living Building Challenge certification in 2015; landscape design by Andropogon Associates for Phipps Conservatory and Botanical Gardens, 2012, Pittsburgh

The Living Building Challenge is 'the world's most rigorous proven performance standard for buildings'.[38] Run by the International Living Future Institute and first created in 2006, the Challenge is now in its fourth version and focuses on restorative standards. This means that projects must contribute more than they take away and have a net positive impact.

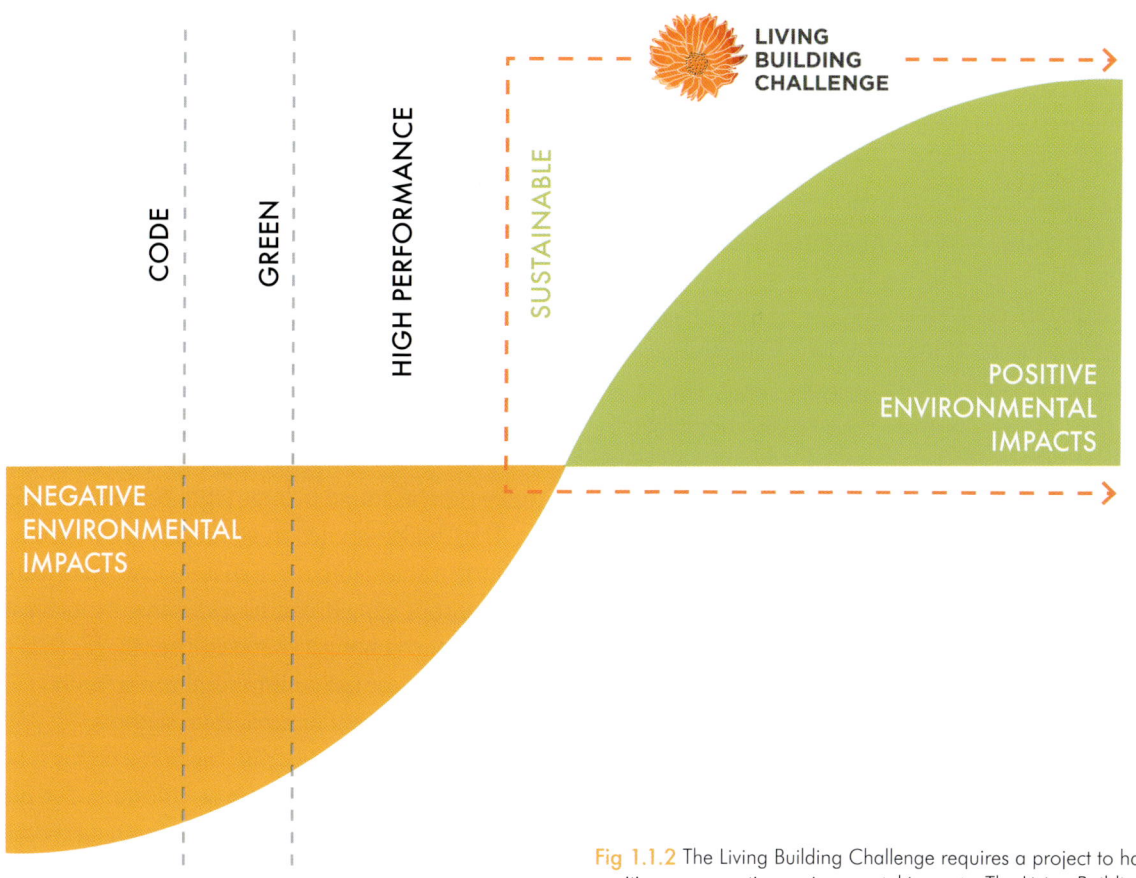

CODE

GREEN

HIGH PERFORMANCE

SUSTAINABLE

LIVING BUILDING CHALLENGE

POSITIVE ENVIRONMENTAL IMPACTS

NEGATIVE ENVIRONMENTAL IMPACTS

Fig 1.1.2 The Living Building Challenge requires a project to have positive, regenerative environmental impacts, The Living Building Challenge 4.0, April 2019

LIVING BUILDING CHALLENGE

Fig 1.1.3 The Living Building Challenge logo

Projects have to operate for a year before they are audited to gain certification, ensuring that there is no gap between the predicted performance and reality. Projects must demonstrate that they can provide 105% of the site's energy needs from renewable resources, excluding any combustible sources, and that all water used is captured on site.

The Challenge is of interest to landscape architects as landscape and infrastructure projects are eligible, as one of the four typologies, along with new buildings, existing buildings and interiors. The landscape and infrastructure typology includes parks, roads, bridges and plazas.

Rather than achieve points via a checklist the Challenge requires a project to meet the standards set by up to 20 different Imperatives, with 14 of these required for landscape and infrastructure projects. The Imperatives are grouped into themes, described as Petals, linking to the Living Building Challenge philosophy that projects should function as 'elegantly and efficiently as a flower'.

Projects must fulfil all relevant imperatives to achieve Living Certification, but they can also achieve Petal Certification if they want to focus on just one of the issue areas or Petals.

The requirements for each project are adjusted depending on the location, based on the New Urban Transect model devised by Duany Plater-Zyberk and Company which mimics the idea of an ecological transect but classifies based on land use, ranging through six zone types from natural to urban core.

For the Living Building Challenge the zones are:

L1 NATURAL HABITAT PRESERVE – this may not be developed except in limited circumstances related to the preservation of interpretation of the landscape

L2 RURAL ZONE – primarily land used for agriculture, or outlying areas of small towns or villages

L3 VILLAGE OR CAMPUS ZONE – low-density mixed use

L4 GENERAL URBAN ZONE – light- to medium-density mixed use

L5 URBAN CENTRE ZONE – medium- to high-density mixed use

L6 URBAN CORE ZONE – high- to very high-density mixed use, as found in large cities or metropolises.

The Challenge encourages areas to move from suburban zones into more sustainable types, rather than encouraging urban sprawl, either by increasing density or changing into mixed-use villages that support low car use, or into rural zones for food production, habitat or ecosystem services.

The Challenge is described as first a philosophy, then an advocacy and then a certification. The projects are meant to be role models to show what is possible, demonstrating true sustainability and working to restore a site.

An interesting Imperative is Beauty – recognising that enjoyment of a space, and that beauty in itself is a positive, sets the Challenge apart from other standards. This is one of the reasons I support the Challenge and feel it is of direct relevance to landscape architecture. That one of the central tenets of our work, the creation of beautiful spaces, is given an equal status to energy efficiency or water use demonstrates the relevance of the Challenge to our sector.

Fig 1.1.4 A lagoon beside Phipps Center for Sustainable Landscapes captures rainwater, contributing to the project's net-zero-water operating standard, while providing a habitat for native fish and amphibians; landscape design by Andropogon Associates for Phipps Conservatory and Botanical Gardens, 2012, Pittsburgh

Fig 1.1.5 (Pages 26 and 27) The landscape and atrium of Phipps Center for Sustainable Landscapes are accessible to the garden's half a million annual guests; landscape design by Andropogon Associates for Phipps Conservatory and Botanical Gardens, 2012, Pittsburgh

Fig 1.1.6 (Page 27) The rainwater capture lagoon and boardwalk at Phipps Center for Sustainable Landscapes are bordered by a winding path of native plant communities; landscape design by Andropogon Associates for Phipps Conservatory and Botanical Gardens, 2017, Pittsburgh

Fig 1.1.7 (Page 27) Cuerden Valley Park Visitor's Centre – currently seeking Living Building Challenge certification, Cuerden Valley Park Trust 2018, Bamber Bridge, Preston, Lancashire

Fig 1.1.8 (Page 27) Cuerden Valley Park Visitor's Centre – currently seeking Living Building Challenge certification, Cuerden Valley Park Trust 2018, Bamber Bridge, Preston, Lancashire

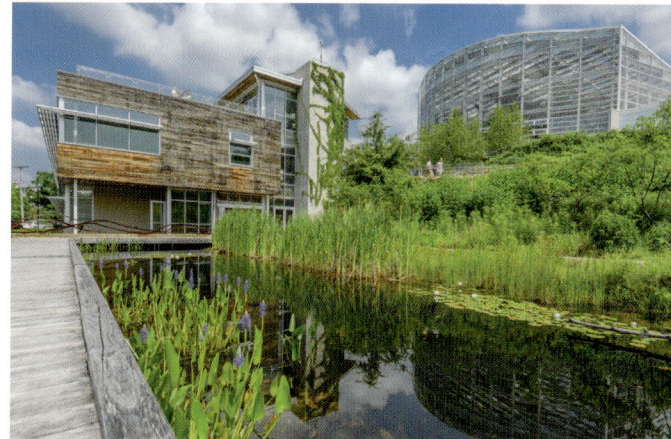

PETAL	IMPERATIVE	LANDSCAPE + INFRASTRUCTURE	INTENT
PLACE Restoring a healthy relationship between nature, place and community	1 Ecology of place	●	to protect wild and ecologically significant places and encourage ecological regeneration and enhanced function of the communities and places
	2 Urban agriculture	●	to integrate opportunities for connecting the community to locally grown fresh food
	3 Habitat exchange	●	to protect land for other species as more and more land is taken for human use
	4 Human scaled living	●	to contribute toward the creation of walkable, pedestrian-oriented communities that reduce the use of fossil fuel vehicles
WATER Creating developments that operate within the water balance of a given place and climate	5 Responsible water use	●	to encourage projects to treat water like a precious resource, minimising waste and the use of potable water, while avoiding downstream impacts and pollution
	6 Net positive water	●	for project water use and release to work in harmony with the natural water flows of the site and its surroundings
ENERGY Relying on renewable resources	7 Energy + carbon reduction	●	to treat energy as a precious resource and minimise energy related carbon emissions that contribute to climate change
	8 Net positive energy	●	to foster the development and use of carbon-free renewable energy resources while avoiding the negative impacts of fossil fuel use, primarily the emissions that contribute to global climate change
HEALTH + HAPPINESS fostering environments that optimise physical and psychological health and wellbeing	9 Healthy interior design	●	to promote good indoor air quality and a healthy interior environment for project occupants
	10 Healthy interior performance	●	to demonstrate ongoing high-quality indoor air and a healthy indoor environment
	11 Access to nature	●	to provide opportunities for project occupants to directly connect to nature, and to assess the success of the Health + Happiness Imperatives

PETAL	IMPERATIVE	LANDSCAPE + INFRASTRUCTURE	INTENT
MATERIALS Building with products that are safe for all species through time	12 Responsible materials	🟠	to set a baseline for transparency, sustainable extraction, support of local industry and waste diversion for all projects
	13 Red list	🟠	to foster a transparent materials economy free of toxins and harmful chemicals
	14 Responsible sourcing	🟠	to support sustainable extraction of materials and transparent labelling of products
	15 Living economy sourcing	🟠	to foster local communities and businesses, while minimising transportation impacts
	16 Net positive waste	🟠	to integrate waste reduction into all phases of projects and to encourage imaginative reuse of salvaged 'waste' materials
EQUITY Supporting a just, equitable world	17 Universal access	◐	to allow equitable access to, and protections from, any negative impacts resulting from the development of Living Building projects
	18 Inclusion	🟠	to help create stable, safe and high-paying job opportunities for people in the local community, and support local diverse businesses through hiring, purchasing and workforce development practices
BEAUTY Celebrating design that uplifts the human spirit	19 Beauty + biophilia	🟠	to connect teams and occupants with the benefits of biophilia and incorporate meaningful biophilic design elements into the project
	20 Education + inspiration	🟠	to provide educational materials about the operation and performance of the project to the occupants and the public in order to share successful solutions and catalyse broader change

🟠 Required   ◐ Requirement depending on scope   ⚪ Not required for typology

Table 1.1.1 Petal certification

Chapter Two
PLAN

INTRODUCTION

Once a project has been instigated, a brief needs to be prepared and the work planned. We need to understand why the client has begun this project. We need to ensure that the whole project team is clear about the project objectives. This sounds an obvious step, but if a project has a tight deadline or if the client is inexperienced this stage can be overlooked. This is a mistake – if issues such as feasibility, project objectives, budget, existing site conditions and sustainability aspirations are not discussed at the outset the client's aims might not be met, and the project team could end up working towards differing outcomes.

In this chapter we will look at the information we need to collate, the questions we need to ask and the decisions we need to make at project inception to set the basis for project success.

THE LANDSCAPE ARCHITECT

Ideally the landscape architect would be appointed at this early stage, allowing us to work on the project brief and to help set the project objectives. This isn't always the reality and we can be appointed at a stage where we have limited influence or an incomplete understanding of how the design has previously evolved. This is frustrating as well-informed decisions during this workstage often result in a better landscape scheme.

Fig 2.0 (Pages 30 and 31) Burnt House Mill, Haddiscoe, Norfolk, 2014

Setting up a good client relationship at this early stage can set the tone for the whole project. If a client does not see us as their equal or has a poor perception of our industry it can be hard to build a successful working relationship.

Is the project worth bidding for?

The amount of unpaid, upfront work required to bid for projects can vary substantially, so it is worth reviewing the procurement option the client has chosen before deciding to submit a bid or fee quote.

The time and costs involved in preparing bids need to be covered, either by the fees from that project or other projects, and projects with a low chance of appointment or an extensive bid process may not be economically viable. We might make the decision to bid for a project for reasons other than profit, such as raising our profile or moving into a new sector, but decisions need to be made based on a clear understanding of the risks involved – unprofitable organisations are inevitably short lived. Useful publications that explore practice management and finance, including an explanation of profit, are listed in Further Reading.

Recording the amount of time spent on bid writing means it can be monitored as a business cost. Detailed recording of the time taken to complete past bids provides a useful reference when deciding whether to bid for similar work. Time-recording software is often included as part of business accounting software, or practices can use web-browser-based software that links to other systems such as project management or capacity forecasting tools.

STAGE 0 – STRATEGIC DEFINITION

- Site selection
- Identify constraints
- Assess landscape character, Landscape and Visual Impact Assessment (LVIA)
- Visit site
- Identify business case requirements
- Undertake an appraisal of the project against national and local planning policy
- Identify end users/stakeholders and undertake a review of grant funding sources
- Identify strategic brief
- Contribute to preparation of strategic brief
- Identify core project requirements
- Agree scope of services
- Establish project programme
- Identify project team
- Project team meeting
- Design team meeting
- Collate feedback from previous projects
- Determine site boundaries
- Determine whether project requires a strategic environmental assessment (SEA) and/or environmental impact assessment
- Carry out a landscape character assessment to inform the baseline
- Carry out a seascape character assessment
- Identify landscape and visual constraints
- Coordinate and input into SEA
- Discuss project with appropriate planning authority
- Identify any existing green infrastructure strategies, policies or plans
- Prepare for stakeholder engagement
- Review published sensitivity/capacity studies
- Comment on project programme
- Test the robustness of the strategic brief
- Review feedback from previous projects
- Ensure that a strategic sustainability review of client needs and potential sites has been carried out, including reuse of existing facilities, building components or materials
- Strategic brief information exchange

STAGE 1 – PREPARATION AND BRIEF

- Landscape appraisal – topography, tree surveys, geology, archaeology, ecology/habitat, vegetation, water-sensitive design, landscape context
- Agree design team
- Contribute to development of initial project brief including project objectives, quality objectives, project outcomes, sustainability aspirations, project budget and other parameters or constraints
- Develop initial project brief with project team including project objectives, quality objectives, project outcomes, sustainability aspirations, project budget and other parameters or constraints
- Project team meeting
- Collate comments and facilitate workshops to develop initial project brief
- Prepare project roles table and contractual tree and continue assembling and appointing project team members
- Prepare schedule of services and develop design responsibility matrix, including information exchanges with lead designer
- Prepare landscape appraisal: topography, geology, archaeology, ecology/habitat, vegetation, water-sensitive design, landscape context
- Review project programme and feasibility studies
- Prepare risk assessments
- Provide information for and contribute to contents of project execution plan as required
- Prepare handover strategy
- Monitor and review progress and performance of project team
- Agree schedule of services
- Agree design responsibility matrix
- Agree sustainability targets
- Agree environmental requirements
- Carry out early stage surveys and/or monitoring
- Assess statutory requirements
- Gather baseline Information
- Site Waste Management Plan
- Design team meeting
- Finalise design brief
- Exchange initial project brief.
- Prepare outline/performance specification
- Confirm handover strategy

Table 2.1 The elements of this workstage[1]

How do I want to be appointed?

How we are appointed can determine our level of influence within the project team. Complex sub-consultancy arrangements, common in the construction sector, can be cumbersome to manage and mean that consultants are far removed from the client, potentially reliant on intermediaries for instructions. Confused lines of communication can also lead to errors and time delays as requests are moved through each level. Good communication within the project team, such as using a Common Data Environment (CDE), a digital location for storing, managing and distributing information, can help track decisions, provide a single source of information and ensure that all involved know the status of all outstanding issues.

Factors to consider include:
- Who is lead consultant – could this be the landscape architect?
- Will we have any influence with the client?

- If the project team share files by email, there are numerous communication routes.
- Data is only visible to the parties in the email.
- It is difficult to search or to control versions.

- If the project team share information via a common data environment there is only one communication route.
- Data is visible to all relevant parties It is easy to search or to control versions.

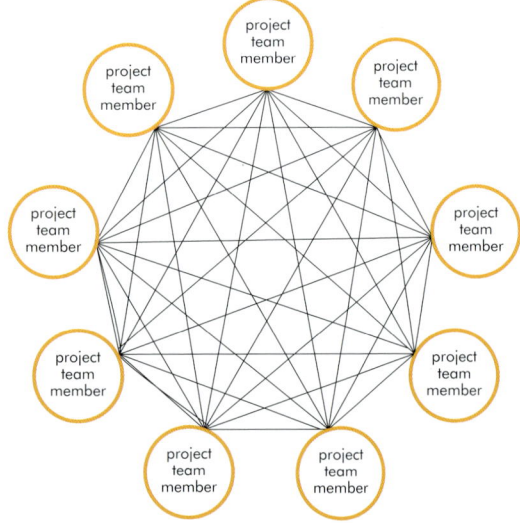
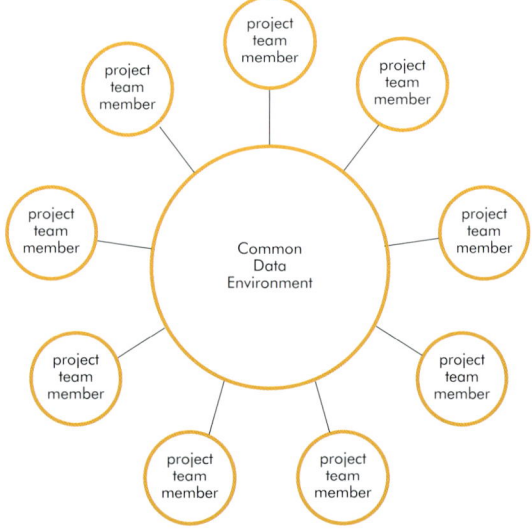

Fig 2.1 Comparison of communication routes in BIM and traditional projects

METHOD	POSITIVES	NEGATIVES
Direct appointment	– Shorter procurement period – No competition	– Project aims or scope of services might not be defined if tender documents are not produced – Need to be known by client
Existing relationship	– You already understand the client – Shorter procurement period	– Can be difficult to raise rates with long-term clients if not re-bidding for work – Production of a project-specific brief can be overlooked
Framework agreement with further competition for specific projects	– Little or no competition once appointed to framework – Build long-term relationship with client	– Longer procurement period – No guarantee of work – Can be tied to rates for a number of years – Requirements such as insurance rates can mean only large practice eligible to bid
Competitive fee bid or financial tender	– Don't need to be known to client	– Lower chance of appointment as number of tenderers is unlimited
Invited competitive tender	– Greater chance of appointment as number of tenderers is limited	– Need to be known to the client to be included
Expression of interest/PQQ only (no design work)	– Simple to complete as no costings required – Unsuitable tenderers are quickly ruled out	– Longer procurement process if part of a two-stage process
Expression of interest/PQQ followed by design competition	– Unsuitable tenderers are quickly ruled out	– Can involve substantial amount of unpaid work
Invited design competition	– Need to be known by the client	– Need to be known to the client to be included
Open design competition	– Don't need to be known to client – Useful for those with a limited track record – Good promotional opportunity even if not appointed	– Low chance of appointment, especially if international competition – Can involve substantial amount of unpaid work
Be the client (eg research bids, speculative projects)	– No procurement process – Can set our own timescales – Can decide focus of the work, such as area of personal interest	– The majority of risk is with the landscape architect

Table 2.2 Methods of appointment

- Do we have a direct contractual relationship with the client or are we a sub-consultant?
- Is our success, or the client's perception of our success, reliant on the performance of others?

New ways of working

The idea of the 'gig economy' has developed rapidly in the last few years, allowing new ways of working based on short-term contracts or freelance work paid per task or 'gig'. App-based working is a more complex version of a taxi rank, where each job is allocated to any one of a pool of registered workers, depending on their location and availability. Two of the best-known examples of this way of working are Uber and Deliveroo. We might think that this working method is far removed from our own but there are parallels, as unless we create our own project we work like taxi drivers waiting for the next project to be allocated, turning our light off when we aren't available for work.

Unlike an Uber driver our appointment process is far slower than a tap on a mobile phone to accept a job, but our clients do need to know our availability, pricing and if the project location is suitable. The digital innovations predicted by Sir Michael Latham in his 1994 construction industry report have similarities with the online marketplaces used to manage independent workers, allowing real time contact with potential clients. The online site Freelancer.com already has a category for landscape design, with regular requests for landscape design tasks. Media coverage tends to focus on the low-skilled side of the gig economy but according to a 2018 McKinsey report knowledge-intensive industries and creative occupations are the largest growing sectors of the freelance economy, with up to 162 million people in Europe and the United States undertaking some form of independent work.[2]

There are serious issues with this way of working, such as a lack of clarity over employment rights, the precarious nature of working, and the potential for discrimination within the allocation process.[3] However, there are concepts within the process that directly relate to our way of working and could become part of our future working practice. To allow innovation in appointment we may need to become more open about such issues as our availability or hourly rates.

New appointment models could be created which share project risk and reward more equally across the project team or improving collaboration, but it is a subject area that sadly hasn't been given much attention. Better appointment methods could be a way to improve quality within our sector or provide better value for clients. Sometimes we have no choice of how a project is appointed, such as a large framework contract issued by a public-sector organisation where the terms are dictated by regulation. Whichever method we chose, or have chosen for us, we need to be certain we understand the potential risks and be certain that they are acceptable.

Fig 2.2 (Page 37) Grafham Water, a manmade reservoir with valuable areas of wetland habitat, was one of the first reservoirs to be designed with a conservation approach, 1962 to 1966, Grafham, Cambridgeshire

The landscape consultant's appointment

In the UK the Landscape Institute publishes a standard form of agreement for appointing landscape architects. The Landscape Consultant's Appointment, first published in 1988, sets out the terms and conditions that need to be considered, included payment terms.[4]

Making money

Whatever standards we have defined for ourselves and our practice, it is still work and it needs to provide us with an income, whether that is directly via client fees or indirectly such as in the internal recharge system used in the public sector. For design professions the idea of profit can be uncomfortable but it has to be considered and shouldn't be something to shy away from. Profit means we can invest in training and technology or build up cash reserves to help counter recessions. Not making a profit means we only just cover our costs, or ultimately have no work. Improved profit can be found by reducing costs as well as by increasing fees – one of the potential benefits of BIM is efficiencies in the time taken to produce our work.

The construction sector has some of the lowest levels of productivity of any sector.[5] Our sector has also been slow to take on the innovations and efficiencies that technology can provide – the McKinsey report shows that in the US only hunting and agriculture made less use of digital innovation.[6] Whilst the use of improved technology doesn't guarantee an improvement in productivity, it is recognised as one of the areas that can create the greatest gain. We should continually work to find efficiencies, ensuring that some of the savings remain with us, as well as being passed on to the client.

Good financial management is central to making money. Again this may sound obvious but not every practice has systems in place to check that projects are within budget. The construction sector in the UK has a track record of low profit margins, with some time periods showing the average as a loss across the whole sector. This is unsustainable and prevents our sector from innovation. The collapse of UK company Carillion proved that having a large portfolio of projects is no guarantee of success.

What else can the landscape architect offer?

A frustration for landscape architects can be the limited scope they are given within a project team. This can be partially dealt with if we are appointed early and involved in the initial decisions, but we can also explain to clients and the project team other ways our skills can be used. Not all clients realise that their landscape architect could add more to a scheme than just complying with planning requirements. Examples of the potential scope for a landscape architect include:

Carbon sequestration – Working to retain the health of existing soil on site and designing planting and maintenance to maximise carbon storage. Healthy soil can store more carbon than neglected soil, as well as providing water storage.[7]

Sustainable site management – Designing schemes without irrigation, or avoiding maintenance techniques that require high energy use (see Chapter 5).

Design for accessibility – Ensuring sites work for all users (see Chapter 3).

Improving financial viability and profit – suggesting creative changes to the design that maximise the

GRASS SEED MIXES TO SEQUESTER CARBON

In 2005 the team at the Top Green Breeding and Research Station in Les Alleuds, France began a research programme to look at the differences in the carbon sequestration values of managed amenity grass species and varieties.

Amenity grassland, sometimes described as a 'green desert' due to the poor habitat value, does provide some carbon sequestration – it is estimated that 2 billion tonnes of CO_2 is stored under UK grasslands.[8] Areas of grass supply long-term storage, locking in carbon predominantly in the soil whilst the grass is in place. However this benefit is often countered by the adverse environmental impact of lawn care, with intensive mowing using fossil-fuel-powered mowers, the application of fertiliser, herbicide and pesticides, the homogenous habitat and the need for irrigation.

Using 1m² plots the researchers tested different grass species, and then different grass varieties, under identical conditions, finding significant differences in the ability of different species to store and sequester carbon within the leaves, roots and soil.

The results of a sequence of trials were impressive. Varieties within the field trail plots:
- sequestered 300% more atmospheric carbon dioxide
- produced 45% less grass clippings
- were wear tolerant.[9]

The levels of carbon sequestration were 13 tonnes/hectare/annum of atmospheric carbon dioxide within topsoil, over 7 times the 2 tonnes/hectare/annum for deciduous woodland[10] and equal to just over two years' CO_2 for the average person in the UK.[11] The deep rooting mixes have good drought resilience, reducing the need for irrigation, as well as increasing rainwater infiltration, reducing surface water run-off and improving the performance of Sustainable Drainage Systems (SuDS).

The reduction in mowing costs and time, the savings in fuel usage and emissions and the reduction in the volume of green waste result in a reduced environmental impact and require significantly less maintenance. The grass seed mixes have been further developed by Rigby Taylor Ltd, the UK importer for Top Green grasses, and are now sold in the UK as Carbon4Grass™ and Carbon Grass™ mixes. The level of carbon sequestration will depend on soil types and local climatic conditions.

In July 2019 the UK Landscape Institute declared a climate change and biological diversity emergency, in recognition of the findings of the Intergovernmental Panel on Climate Change (IPCC) and in support of the UK government commitment to reduce net carbon emissions to zero by 2050. Given there is a negligible difference in price between the carbon sequestration mixes and traditional amenity seed mixes, and that existing grassland can be over-seeded with minimal disruption to use, reviewing the grass mixes we specify could be an important part of our work as landscape architects to mitigate the impact of climate change.

Fig 2.3 Trial plots to assess the carbon sequestration characteristic of amenity turf, 2016, DLF Trifolium Plant Breeding Centre, Les Alleuds, France

commercial benefits, such as more housing or more rock extraction, without compromising other factors.

Improvements in air quality – Trees take up air pollutants, including ozone and nitrogen oxides, as well as reducing the fine particulate matter in air pollution that can damage health.

Localised climate adaptation – Tree canopies, structures, water features and choice of materials can help reduce the air temperature in hot weather or reduce the impact of cold weather.

Even amongst other construction design professions there is a perception that designed landscape is only the soft landscaping, just the green fluffy stuff. We need to ensure that the whole of the designed landscape, from the kerbs to the paving to the street furniture, is recognised as part of the scope of our work, and that we are involved in the design of these elements. A well-designed landscape can be almost invisible, as it doesn't interrupt our passage through a space and looks in keeping with the setting. This may be why our work is sometimes overlooked – if we do our job well no one notices we have been there.

THE CLIENT

As we recognised in Chapter 1, without a client we have no project and with no projects we have no practice. Knowing the standards that a potential client expects and aspires to is an important part of project inception – if these are mismatched then either the client could be disappointed or we deliver more than is expected by the client for no real benefit.

Fig 2.4 Artificial insect habitat made from clay pipes, 2014, Oxford Botanic Garden

The client – definition and issues

Without clients we have no work. The project is a response to their need, and it is our job to define that need, determine if we are the right person to address it, and if so develop a solution. In some cases the client role is clear – the person or organisation who instigates the project, decides the parameters and pays the bills is the client. In other situations it can be less clear – a public park restoration might be on a site owned by a local authority, funded by a charity and used by the public. Defining roles and responsibilities at the start of the project, including who takes the ultimate client role, is an important part of defining a project. Roles and responsibilities are explored later in this chapter.

With complex sub-consultancy arrangements we may have a direct client or agent of the client who commissions our work and pays our fee, and an ultimate client who has instigated the project and makes the final decisions. Agreeing how decisions and changes are approved at the outset of a project can prevent confusion as work progresses. Receiving conflicting instructions from multiple sources is a frustrating and unproductive way to work. Our client might in reality be a group of individuals but a process needs to be agreed at the outset, so that decisions are made by those with a full understanding of the project and the consequences of any changes.

Defining our relationship with the client is an important part of project inception. They need to know our working methods and standards, and we theirs. How quickly will we reply to emails? Will we be contactable out of office hours? How will additional work be agreed? Is the client happy to share aspects of the project on social media? How will data be managed?

Given that clients are vital to most of our projects there is surprisingly limited research into the client–designer relationship. The Latham Report, commissioned by the UK government and industry bodies and published in July 1994 by Sir Michael Latham, highlighted the shortcomings of the relationship, but minimal guidance on how to manage the relationship has been produced.[12]

Whoever the client is, they own the project. It is not ours and doesn't provide an opportunity for us to express our taste at the expense of the client's wishes. The design must function, meet the requirements of any relevant legislation, demonstrate good design and be fit for purpose, but the rest of the options are the choice of our client with guidance from us. A true test of any designer is to deliver a scheme that performs as a landscape but in a style at odds with their ideas. A detachment of personal views can be difficult, especially if we are hoping to use the project to promote our work, or we are worried that it will impact on our reputation. This is a separate issue to standards or ethics – the client is not asking us to break the law or behave unprofessionally but where the client's views clash with our own it can still be an uncomfortable situation to manage.

It is important to remember that the client may have concealed needs that they never express, such as a wish to be promoted, or to avoid losing their job. In the construction sector our work is about change, whereas for many sectors the focus is on preventing change.

Change can cause uncertainty and emotion – we need to remember that for us the project might be a career-defining experience but for our client it could be an uncomfortable intrusion into their normal working day. We ask a client to trust that we have correctly understood their needs, and that we will be able to translate these ideas to deliver at a future time.

What do clients want?

Part of our role is to explain all relevant issues to our clients, and work to create a clear brief. It is also our role to define our expectations of the relationship. This is partly set out in the contract, but it is also determined by our own conduct.

We can't predict how the client–designer relationship will work but there are several factors to consider when deciding to accept a project, and what level of fees to quote. We can never fully understand all our client's wants, particularly as they may not fully understand them themselves, but we need to discover as much as possible at this stage.

Careful questioning can help uncover some of the client's requirements, as well as taking the time to understand how their project fits with their day-to-day work. Using a mix of open and closed, funnel, probing and leading questions can help reveal as much as possible about the client's requirements, including helping them develop and refine those needs or consider other options.

The diagram (left) shows some of the factors to consider when assessing a potential client.

The 5 Whys technique, developed in the 1930s by Toyota founder Sakichi Toyada, can be a useful way to make sure we have a complete understanding of an issue, and not just the superficial symptom.[13] A deceptively simple technique, it involves following up the first answer with another 'why?' Five is an arbitrary number – you ask why until the root of the issue is revealed.

Assessing the client

Some of the warning signs of potentially bad clients can include unrealistic timescales or budgets, only appointing professionals when forced to by third parties such as planning authorities, or trying to pay below market-rate fees. Defining clients as good or bad is simplistic – rarely is the situation so clear cut. With hindsight we might begin to spot client traits that tend to produce better projects, but we also need

to look at our own role, how we communicate and how we affect the outcome. How someone approaches commissioning a landscape architect may give some clues, but we can never predict how the client–designer relationship will develop. Sharing similar values to those of our client can help, but it is equally possible to run successful projects for clients with differing views from our own.

The client–designer relationship was a challenge for Lancelot 'Capability' Brown and will probably remain

LOW RISK ⟶	HIGH RISK
Client role is central to their day job	Client role is additional to their day-to-day work
Well informed	Uninformed (naive) or partially informed client
Frequent client	First-time or infrequent client
Is fully supported by their company or organisation	Isn't fully supported by their company or organisation
Appoints project team as early as possible	Appoints project team late in the design process
Realistic budgets and timescales	Unrealistic budgets and timescales
Supportive of project	Reluctantly acting as client against their personal views

Table 2.3 Client assessment (based on the work of Boyd and Chinyio)[14]

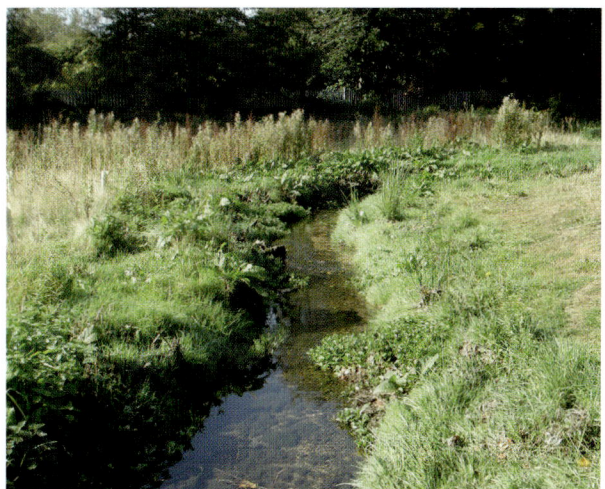

Fig 2.5 Flood alleviation scheme – the business case for publicly funded projects is made using a methodology set out by central government; scheme funded by London Borough of Harrow and Environment Agency Pinner in 2008, Greater London

a potential challenge for future designers. However, it is incredibly rewarding to take a client's germ of an idea and work with them to create a scheme that meets their need.

'Mr Brown could not get the money for the Extra Work and tore the account before Mr Dickens face and said his say upon that Business to him.'

Extract from 'The Account Book of Lancelot "Capability" Brown, the great landscape gardener of Fenstanton Hants',[15] describing a dispute in 1765 between Brown and his client Ambrose Dickens of Branches, Suffolk over extra work worth £58 1s 8d. In 1764 his gross income was £6,000, so it was not an insubstantial sum.[16]

The business case

Part of this workstage is to understand the business case for the project – this is a broad term that doesn't only apply to commercial sites but is relevant to most landscape schemes. Why has the client decided to go ahead with this project now? How will it be funded? What approvals are needed to move to each step in the process? What are the impacts of a Do Nothing option? These are all questions that should be answered in a clear business case.

Publicly funded projects, such as flood alleviation schemes or new parks, will usually have been assessed against a set of criteria to demonstrate their value. The criteria may be financial (homes saved from flooding, reduced impact on transport) but they can be social (improved health and wellbeing) or environmental (improved air quality or the creation of new habitats). Understanding why the project has been instigated and what factors where part of this decision can help ensure that an accurate brief is written, and in turn a project is delivered that meets the client's requirements.

Landscape as a process

Landscape architecture is an ongoing process, not an outcome, and our role is to help clients understand that. We work to change how an area of land appears, and that change will be ongoing. Unlike the rest of the construction sector we design using living materials which have natural variation and need care to survive. The passing of time is a central component of a landscape scheme – as trees and plants mature views will change and the ultimate concept for the site will be realised. In contrast buildings are static – they may weather or age but they do not change substantially without human intervention.

Good clients understand that landscape is a process that they need to support, and have the patience to allow them to develop. Almost instant landscapes are possible with enough money but even the largest trees for sale are dwarfed by fully grown versions. Part of our work is to explain that landscapes work in four dimensions, and explore how a client's need for impact might be balanced with long-term appearance. The landscape scheme may long outlive any associated buildings, as well as the client and the design team.

Engaged clients

Over a long timescale it can be hard for a client to retain enthusiasm, so part of our role is to keep the client engaged in the process, supporting them through any difficult stages. It can be useful to forewarn clients of any workstages that you expect to be problematic or where progress will appear minimal. It is also important to discuss with clients how much time you will need them to give to the project, and agree a realistic response time from them for queries. The nature of the project will determine the level of client involvement, as will the client's level of interest in the design. If the client has limited availability or is out of contact for long periods of time this needs to be recognised in the project brief and the project timetable.

Having an engaged client, who is interested in the details of their landscape scheme, who has a long-term commitment for caring to and maintaining the site and who is keen to explore new ideas is a privilege that should be valued.

Budget

Getting the client to reveal their budget can be difficult – some clients seem to feel that if they reveal their budget they may not get value for money. Other clients may say they don't know their budget, but even at the earliest stages of the project they will have an idea of the most they would commit to the project, even if the price bracket is broad. We need to know at least the approximate budget as early as possible so that we can match our fee quote to the work required. We might be imagining an extensive new design with high-quality materials whilst the client is imagining a more basic design.

If the budget isn't enough to meet the client's aspirations this needs to be addressed now. Most clients have a mismatch between their aspirations and their budget. It is possible to work with the client to adjust the scope of the project to reduce costs, or to persuade them of the value of investing more in their landscape scheme. However, allowing a client to continue with an unrealistic budget is unfair to the client and is certain to lead to an unsuccessful project outcome.

Fig 2.6 gives a stark illustration of the trend in construction to allow clients to believe their budget is adequate and then expect them to cover the gap between the estimate and the reality. In a presentation in 2018 by Mark Farmer, discussing his 2016 report 'Modernise or Die! – Time to decide the industry's future' he suggested that the increase in variance in times of recession shows the deliberate practice of underestimating costs to win work.[17] As his report illustrates there are other factors involved in causing this difference, such as clients changing their requirements at a late stage or the lack of a well-defined brief, but optimistic (or possibly cynical) estimating must be accepted as part of the reason for this mismatch.

THE PROJECT

At this stage the only details we have for the project are the specific need that we have helped the client define, and the approximate location where that need will be solved. There may even still be a choice of locations if the client is involving the project team in their feasibility assessment. The budget may give us an indication of scope or scale but there are still numerous parameters left to define.

To create a good brief for a project we need to understand exactly who has a stake in the project. Who

TENDER PRICE INDEX VS CONSTRUCTION COST INDEX (1985=100)

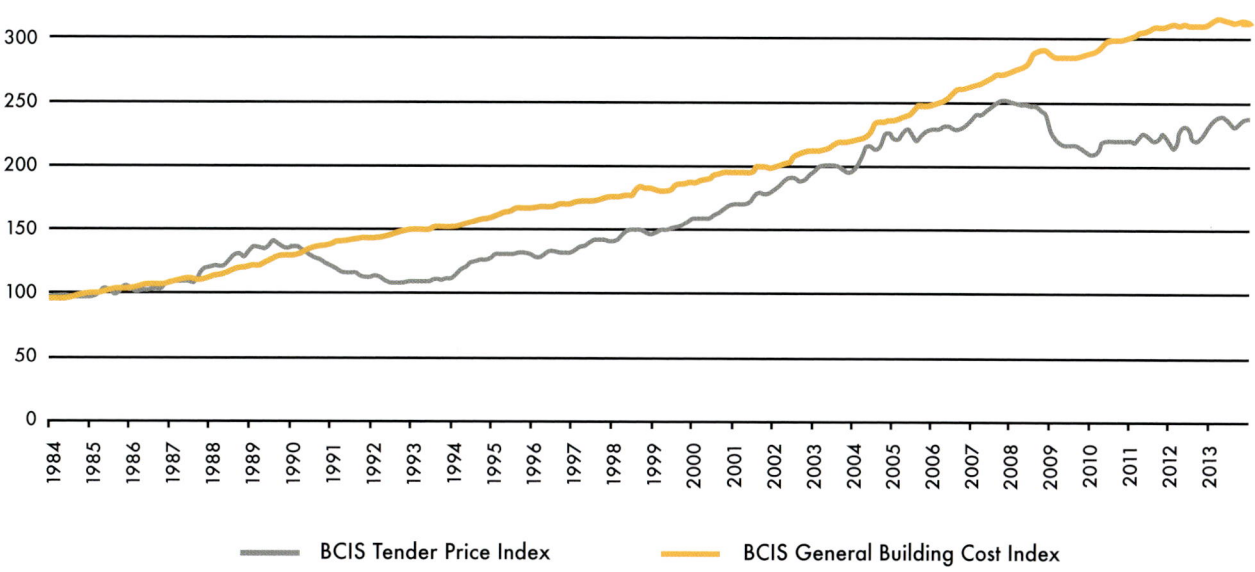

Fig 2.6 Difference between quoted and final price in construction project - based on the work of Mark Farmer using BCIS indices

cares about the site? Who will be affected by the work, either by the completed scheme or during construction? Even sites with apparently low landscape value can be important to someone, especially if linked to family memories. We need to be careful not to be dismissive of any landscape. To us it might be an unused site in urgent need of redesign, but for others it could be the place they feed the ducks with their family or visited as a child. It could be the only green space available to those in the area, or a valued view for someone who is housebound.

It is important to take time to really understand a site. We need to see it at different times of day and in different weathers, as a short visit may not be representative. We need to know about irregular uses, such as public events. Events such as bonfires or festivals have requirements that we would need to accommodate in our design, such as higher levels of emergency access or temporary parking. We can use online information, such as photographs posted on social media, news articles or air photography, to gain some understanding of a site but nothing beats a series of site visits.

Site survey and assessment tools

As well as visiting there are other tools we can use to assess a site. New technology can give us insight that would have been previously impossible or prohibitively expensive.

Techniques available to the landscape architect include:

Open data – The UK government has led the world in the release of government data for public use.[18] Using a geographic information system (GIS) to overlay layers of information can allow analysis of factors as diverse as levels of air pollution, habitat type and provision of

Fig 2.7 A site that appears to have little landscape value may be the place someone fed the ducks as a child

Fig 2.8 Sites may have uses that a short visit won't identify – Christmas at Kew event, 2018, Kew Gardens, London

open space. Free, open-source software such as QGIS can be used to analyse the data and create plans.[19]

Drones – For inaccessible or dangerous sites a drone can be used to view the site, or to allow views to be assessed from set heights or viewpoints. Repeating the same drone flight at different times can be a useful assessment tool, such as recording the nature of a flood or changes to a habitat.

Cloud-point scanning – This allows accurate surveys to be completed in a fraction of the time taken by other methods. The survey can include colour or black and white imagery, so that a virtual model of the site is created. Elements can be measured directly from the model. Scanning equipment is now small enough to be carried by drone, allowing large or remote areas to be rapidly surveyed.

Remote cameras – To supplement site visits a remote camera can be set up to record patterns of site use or assess wildlife. Time-lapse photography created using a remote camera can also be used as an assessment tool, especially if it is in place for the whole of the construction and establishment stages.

Programme and planning

At this early stage the programme will lack detail but it should prove that the client's timescale is feasible, and include any fixed milestones that need to be achieved such as opening dates. It is also a chance to assess capacity within the project team. It is important to discuss how programme changes are agreed and any additional fees approved. A project running longer than planned might allow more time to consider design, but it also incurs more costs such as calls, responding to

emails or site visits. Accepting an extension to a project without increasing fees can erode any profit. Fee quotes should include the expected project duration, and where possible an additional fee per week or month for any overrun outside our control.

If the project requires planning permission or other consents these need to be included on the programme as early as possible, with a realistic timescale for approvals. For some sites it might be appropriate to hold pre-application meetings with the planning authority at this early stage to ensure that the concept is in line with local planning requirements.

Review previous projects

This is the stage to review the site and any similar projects, either with this or other clients, and incorporate any lessons learnt. A system of review needs to be agreed and all relevant information shared. This could include previous planning applications, site investigation reports or habitat surveys.

Feedback and review is now a recognised step in the RIBA Plan of Work, acknowledging that we should learn from others in our sector and avoid repeating known problems. This creates a circular plan of work, with feedback from completed projects informing the early stages of subsequent work.

Defining sustainability standards

If the client wants to achieve a sustainability standard this needs to be decided at this stage. The standard may require the project to be registered, or there may be registration fees that need to be included in the project budget.

Some consultants tell clients the difference in cost between their planned scheme and reaching a recognised sustainability standard, to show that it may be attainable within their budget.

THE PROJECT TEAM

Whether the project team is assembled via an open procurement or a by direct appointment, the risks inherent in setting up a successful project team are similar. All the required skills need to be covered within the team, and that team needs to be able to work together effectively. This sounds a basic requirement but isn't always the reality.

One consideration when assembling a project team is passing on of skills and experience. The 2016 Farmer Review highlighted the loss of experience within the construction sector, with an average four people leaving for every person entering the labour market. As those joining are typically less experienced than those leaving, the net effect is a decline in skills.

There are numerous issues to resolve if we want to halt this decline but giving project experience to those at an early stage in their career is an important factor. They can bring skills or insight that more established professionals may lack. Funding this opportunity can be difficult, especially given the low profit margins in our sector, but if expertise is not passed on valuable knowledge is lost. Having the financial flexibility to allow junior staff to gain experience is another justification for landscape practices to make a profit.

The skills offered within a project team may not be fully evident if an assessment is made by profession alone. Previous careers or interests outside work could provide valuable skills or useful connections in support of the project.

Roles and responsibilities

It is important for all members of the project team to understand the roles and responsibilities of their colleagues. Agreeing which profession leads on the design of each aspect, and who needs to be consulted on any subsequent changes allows changes to be dealt with quickly and for any consequences to be identified. Some aspects will require a collaborative approach, possibly with the structural, aesthetic and environmental issues each dealt with by a different profession.

Given that misunderstandings about the role of the landscape architect within a project team are common, it is likely that we will make similar mistakes with other professions. To work collaboratively we need to appreciate the skills of our colleagues and admit if we don't understand the exact nature of their work, however established we are in our careers. Our understanding of their work could be shaped by our tutors at college, prejudices within our sector or old rivalries between professions, none of which are helpful. Discovering innovations in other professions can help us deliver better landscapes.

For landscape architects it is essential to impress on the project team which aspects of the design we need to be involved in. Some team members might think we don't need to know about underground

services, overlooking the fact that the trees and plants we specify could be impacted by the placement of services.

Forms of contract

Traditional construction projects are hierarchal, with a lead consultant who has direct contact with the client, and potentially a number of tiers of sub-consultancy in their indirect control. This places much of the control with the lead consultant, who is able to decide what the client is told or is asked. Having a single point of contact with the client does avoid duplication but it isn't always ideal.

The alliance contracting model includes the client as part of the project team, with an agreed system of communication so all parties are aware of all outstanding queries and decisions made.

Creating a project team where roles are clearly defined, that includes the range of skills and expertise required, allows younger staff to learn and enables all members to communicate effectively with each other and the client should be our aim. In reality teams can be hurriedly assembled, with gaps in knowledge or duplication of roles overlooked. However a team is put together we need to accommodate any shortcomings and try to mitigate these.

WIDER SOCIETY

Assessing the impact of our work on wider society is an important consideration. As landscape architects our work is likely to long outlive us and we need to consider the impact on future generations. These impacts can be linked to the site, such as changes in the use of the site or visual impact, or wider impacts such as the use of non-renewable resources.

Practical considerations such as noise, traffic or loss of open space during construction need to be considered. Identifying these factors as early as possible may mean that their impact can be reduced. If we can't mitigate we need to decide how we will explain these limitations to the local community, how we will keep them updated and how we will deal with any new issues as they arise.

The 2016 Farmer Review identified that the public has a poor perception of the construction sector. If we want to improve this perception, and in turn encourage new people to enter our sector, we need to ensure that the public experience of our scheme is the best it can be. If we can't meet their expectations we need to clearly communicate the reasons. Every muddy road outside a construction site, every loud radio disturbing a shift worker's sleep and every lorry blocking access adds to the poor perception of our sector. Some annoyances are unavoidable but many can be foreseen. If we want people to support our projects and care for the landscapes we create we need to show equal care whilst creating them.

Planning for wider society

Many of the skills a landscape architect brings to a project have benefits for wider society, as well as achieving benefits for the site itself.

Examples of how our work can benefit wider society include:

- surface water management/Sustainable Urban Drainage Schemes (SuDS)
- low-carbon maintenance methods (see Chapter 6)
- improvements to air quality provided by planting schemes
- habitat creation and restoration
- providing opportunities for improved health and wellbeing

Landscape architects have always recognised the wider benefits of their work but these benefits have begun to be recognised by third parties such as governments and funders. The concept of ecosystem services recognises the contribution of our work, albeit from a human-centric viewpoint, and can be a useful tool when explaining the positive consequences of our work.

Another benefit that our work can provide is improving safety and security for users of public spaces – this is explored in Chapter 3.

The planning stage, before final decisions are made, sets the tone for the rest of the project. It allows the project team to gain as full an understanding as possible of the client's needs, understand the site and to assess the potential impacts on wider society. Time spent at this stage can help reduce the gap between our client's expectations and what is delivered.

At the end of this stage a framework for all future work will have been created and the lessons from previous work consolidated to allow us to move to the next stage.

Fig 2.10 Our work can often inconvenience the public – footpath made almost impassable by excavation work

Case Study 2.1
MAGGIE'S CENTRES
ARCHITECTURAL AND
LANDSCAPE BRIEFS

Title	
Maggie's Centre (Maggie's)	

Client	
Maggie Keswick Jencks Cancer Caring Centres Trust	

Location	**Design Period**
Various – see table 2.1.1	Various
Construction Period	**Type of Scheme**
Various	Therapeutic landscape
Project Value	**Landscape Architect**
£2 million (Maggie's West London)	Various – see table 2.1.1
Owner	**Design Team**
Maggie's Centres	Various – see table 2.1.1

Maggie's Centres are a network of 22 specialist support centres in the UK, Hong Kong and Japan for people with cancer, and their friends and family.[20] Each centre is set in the grounds of a cancer treatment hospital and provides practical, emotional and social support.

Named after Maggie Keswick Jencks, the centres provide a space for counselling, alternative therapies and other activities, as well as being a place to make a cup of tea or find privacy.

Fig 2.1.0 (Page 51) Landscape scheme for Maggie's Dundee, Arabella Lennox-Boyd, 2003, Dundee

Fig 2.1.1 (Page 52) Maggie's Lanarkshire; rankinfraser landscape architecture, Airdrie, North Lanarkshire

Maggie was inspired to create the centres through her own experience. She was diagnosed with breast cancer in 1988 at the age of 47 and was successfully treated. Five years later the cancer returned, now present in her bone, bone marrow and liver, and she was told that it was incurable. Innovative treatments provided a period of remission, but in 1995 the cancer returned for the third time. Whilst her medical needs were met, she felt that there was little practical or emotional support or guidance in wider issues such as nutrition, complementary medicine or support for friends and family.

Her response was *A View from the Front Line*, a description of her experiences of treatment. Her work lead her to the idea of a dedicated centre, with an office and a library. She persuaded the Western General Hospital in Edinburgh, the place she received high-dose chemotherapy and stem-cell replacement, that they needed such a centre. A small former stable in the hospital grounds was chosen as the location and Richard Murphy Architects drew up plans for the conversion with Emma Keswick, the wife of Maggie's cousin, designing the landscape scheme.

Maggie died in 1995, a year before the first Maggie's Centre was built, but her husband Charles Jencks continued to support the project based on her plans.

From one centre more centres followed, all funded by voluntary donations and independent from the hospitals they support. The centres are designed to be less medical and more approachable than the main hospital, and to provide a home from home for those using the building. As a client the centres focus on

Maggie's and its local community

Each Maggie's Centre is unlike any of the others. We need the local community to be proud of their own Maggie's... we need the people who live near them to know that they have somewhere wonderful to turn to should they need to use it. It is "their" Maggie's, it belongs to them and they are proud of it.

We hadn't realised, until it happened, how important this element was for the fundraising that is needed for each Maggie's Centre, as each Centre is self-funded. Both the capital costs of the building in the first-place and then the annual running costs thereafter have to be raised. The building and landscape need to be their own ambassador in their local community. We rely on people knowing and talking about "their" Maggie's.

How the building and garden will be used

Some will visit a Maggie's Centre for the first time when they get diagnosed. Others will not be ready to address the emotional fallout of having cancer, sometimes until long after their treatment. If you have or have had cancer, everything is not necessarily done and dusted by medical treatment.

Families and friends might visit during or after the treatment of someone they love, or perhaps even after someone they love has died. People are very likely to come in for one reason: for example benefits advice or because they are brought in by a friend, and end up using the Centre in a different way to the one they had originally envisaged. With guidance from one of the professional staff who works there, they may then make use of some other part of the carefully tailored programme of support Maggie's offers.

Fig 2.1.2 Extract from Maggie's Architecture and Landscape Brief, 2015

Fig 2.1.3 (Page 55) Maggie's Glasgow courtyard, landscape design by Lily Jencks, 2018, Glasgow

good design, and have commissioned leading landscape architects and architects.

The architectural and landscape briefs for the design of new centres are a good example of a clear brief. The briefs include aspirations for the site as well as specific requirements. The clear instructions for the design of the toilets is typical.

'Toilets: Two toilets with washbasins and mirrors, which should be big enough to take a chair and a bookshelf and one of them must have disabled access. They must be private enough to cry. They must be nice places; they should NEVER have gaps beneath the doors.'[21]
Maggie's Architecture and Landscape Brief 2015

The brief includes specifics, such as including a location to hang coats and store your brolly, but allow enough scope for each site to have a very different look and feel, with the landscape scheme a strong part of the character of the centre.

'Views out: It is important to be able to look out and even step out from as many of the internal spaces as possible even if it is only into a planted courtyard. Planting works well here too. It not only gives a focus to look out at, it can filter privacy in a room with glass doors or windows to the outside. We want the garden, like the kitchen, to be a space for people to share and feel refreshed by.'
Maggie's Architecture and Landscape Brief 2015

The briefs are written in natural language, as if the client is talking directly to the potential designer. This contrasts with many design briefs, where there can be a tendency to write in formal, pseudo-legal language, confusing complexity of design with a need for complex language. Simplicity, clarity and honesty are central to the Maggie's Centre briefs. The parameters are set for the designer to work within. There are also photos of previous projects to show the style and variety of work to date.

'Our buildings and our garden landscapes have to invite you in. The path to the Centre must beckon and guide you to what is clearly the front door. The way the path is planted can help you shed a little of the stress of the hospital atmosphere before you even reach the front door. The landscape gives a bit of breathing space between the two worlds of hospital and normal life (which isn't quite so normal anymore).'

Maggie's Architecture and Landscape Brief 2015

The client's expectations of their role are clearly stated, as is the intention that those involved in the project will enjoy the process.

'Client Team. Maggie's has a small client team and we like to be involved at every stage of the design from the commissioning of the building right through to the opening and beyond. This is a personal not a "committee" project. As clients, we see our job as trying to imagine, at every level, how these buildings will work for the people who will be using them. We want to enjoy ourselves, and for you to do so too. We think we will get a better result if we do. We want to be surprised and delighted. If we are, the people who come to them will be too.'

Maggie's Architecture and Landscape Brief 2015

Fig 2.1.4 Sketch of Maggie's Lanarkshire; Reiach and Hall Architects

TO BE READ IN CONJUNCTION WITH PLANTING SCHEDULE AND SPECIFICATION

Fig 2.1.5 Maggie's Oldham planting plan; Rupert Muldoon for dRMM, 2017, Oldham

Fig 2.1.6 (Pages 58 and 59) Large elm trunks with mirrored steel top, Maggie's Glasgow; design by Lily Jencks in collaboration with Archie McConnel, 2012, Glasgow

Fig 2.1.7 (Page 59) Maggie's Glasgow, view from dining room, landscape design by Lily Jencks, 2018, Glasgow

Fig 2.1.8 (Page 59) Maggie's Highlands; landscape design by Charles Jencks, Page\Park Architects, 2006, Inverness

Fig 2.1.9 (Page 59) Maggie's Lanarkshire, rankinfraser landscape architecture, Airdrie, North Lanarkshire

The quality, diversity and innovation of the Centres suggests that the brief is clear enough to achieve the aims but flexible enough to allow a different interpretation on each site.

	OPENED	LANDSCAPE ARCHITECT	ARCHITECT
Edinburgh	1996	Emma Keswick	Richard Murphy of Richard Murphy Architects
Dundee	2003	Arabella Lennox-Boyd	Frank Gehry
Highlands	2005	Charles Jencks	David Page and Brian Park of Page\Park Architects
Fife	2006	Gross Max Ltd	Dame Zaha Hadid of Zaha Hadid Architects
West London	2008	Dan Pearson Studio	Lord Richard Rogers of Rogers Stirk Harbour + Partners
Cheltenham	2010	Dr Christine Facer	Sir Richard MacCormac of MJP Architects
Glasgow	2011	Lily Jencks	Rem Koolhaas of OMA
Nottingham	2011	Envert Studio	Piers Gough of CZWG Architects
Swansea	2011	Kim Wilkie	Kisho Kurokawa Architects with Garbers & James
Cambridge	2012	Interim centre in existing building	Interim centre
Newcastle	2013	Sarah Price	Ted Cullinan of Cullinan Studio Architects
Hong Kong	2013	Lily Jencks	Frank Gehry
Aberdeen	2013	Snøhetta	Snøhetta
Merseyside	2014	Temporary centre – no landscape scheme	Carmody Groarke
Lanarkshire	2014	rankinfraser landscape architecture	Neil Gillespie of Reiach and Hall Architects
Oxford	2014	Touchstone Collaborations	Chris Wilkinson of Wilkinson Eyre Architects
Cancerkin Centre, London	2016	In existing building	In existing building
Manchester	2016	Dan Pearson Studio	Lord Norman Foster of Foster + Partners
Forth Valley	2017	Darren Hawkes	Garbers & James
Tokyo	2017	—	Tsutomu Abe
Oldham	2017	Rupert Muldoon	Alex de Rijke of dRMM

Table 2.1.1 Maggie's Centres

Case Study 2.2

CLOUD POINT
SCANS OF HISTORIC
BROADS DRAINAGE MILLS

Title
Water Mills and Marshes Landscape Partnership - Cloud point scans of historic Broads drainage mills

Client	Owner
The Broads Authority	Various

Location
Various locations across the Broads National Park

Project Period	Construction Period
2016 to present	2018 to present

Type of Scheme
Historic building conservation in protected landscape

Project Value
£3.957 million over 38 projects (total project value)

Landscape Architect
Claire Thirlwall (acting as Expert Adviser to the National Lottery Heritage Fund)

Project Team
Scheme manager - Will Burchnall, the Broads Authority
Surveyor Anglia Land Surveys Ltd

Main Funder
The National Lottery Heritage Fund

Fig 2.2.0 (Page 61) Halvergate High's Mill, 2014, Halvergate Marshes, the Broads National Park

Fig 2.2.1 (Page 62) Lichen on brickwork, 2018, North Mill, Reedham, the Broads National Park

The project

In 2015 the Broads Authority, the public-sector organisation responsible for the Broads National Park, was awarded £367,000 to develop their bid for a scheme to enhance and conserve the internationally important Broads landscape.

Funded by the National Lottery Heritage Fund as part of their Landscape Partnerships grant programme the Authority worked over a period of 18 months to refine their plans, update costings and test the feasibility of each of the 38 projects within the scheme as well as bid for match funding, build a diverse partnership of supporting organisations and consult with the community. At the end of this development phase a

Fig 2.2.2 Mill workings, 2014, Lockgate Mill, Halvergate Marshes, the Broads National Park

successful Round Two application was submitted and a full grant of £2.437 million awarded, allowing the five-year programme of work to commence.

The landscape

The flat, open landscape of the Broads is characterised by large level fields bounded by ditches rather than hedges, creating an extensive drainage network that now provides valuable wildlife habitat. With few vertical features the distinctive drainage mills, dating from the 18th century and used into the 20th century, are the most prominent features in the landscape.

The complex Broads landscape is in part the result of the flooding of the shallow lakes and lagoons created by medieval peat diggings. This manmade and highly managed landscape has a high conservation value, with 2 RAMSAR sites, 28 Sites of Special Scientific Interest and 17 regionally and locally important wildlife sites.[22] The Norfolk and Suffolk Broads is Britain's largest protected wetland.

The mills were designed to pump water to drain the marshland, allowing farmers to graze the previously unproductive land. Each mill is a valuable snapshot of the technology available at the time, with refinements and innovations added over the period the mills were in use. With no contemporary use many of the mills are in a poor state of repair with some in danger of collapse.

Fig 2.2.3 Cloud point scan of Muttons Mill, 2019, Halvergate Marshes, the Broads National Park

Fig. 2.2.4 Initial output from cloud point scan, Lockgate Mill, 2018, Halvergate Marshes, the Broads National Park

Fig 2.2.5 (Page 65) Cloud point scan showing horizontal section of Lockgate Mill, 2016, Halvergate Marshes, the Broads National Park

Surveying the mills

As the first stage of a programme to conserve 12 of the mills at greatest risk a survey of the structures was commissioned. The original plan had been for the mills to be surveyed manually but following up a suggestion from the National Lottery Heritage Fund adviser the team decided that cloud-point scanning would be a faster, safer and more cost-effective option.

Working from ground level the surveyor used a tripod-mounted laser scanner to take thousands of measurements per second. From the laser scan a 3D digital model was created – this was then used to create accurate base plans for restoration work.

High-resolution colour imagery was also recorded and overlaid onto the 3D model. Using specialist software accurate measurements can be taken from this imagery, allowing the size of components within the buildings to be checked without re-visiting site.

The scanner works anywhere that is has line of sight, so was able to survey areas unsafe for human access. Features were revealed that had been previously hidden as the laser picked up objects through small apertures in broken floors.

The aim of the original surveys was simply to provide accurate base plan information for repair work, but with the data collected via cloud-point scanning other possibilities are now being considered. Digital models of the mills could be shared online, so inaccessible mills can be explored virtually. This could allow a worldwide community of mill enthusiasts to review the workings of each mill and help with the understanding of the previously hidden technology.

Fig. 2.2.6 Cloud point scan in progress at Lockgate Mill, 2016, Halvergate Marshes, the Broads National Park

The mills could be rescanned at regular intervals to assess deterioration and to help prioritise those in most urgent need of repair.

As technology progresses the size of the scanners reduces and the speed of scanning increases – it is now possible to mount them on drones, allowing another perspective to be gained and for the latest advances in technology to help conserve technology from our past.

Fig. 2.2.7 Stones Mill, 2017, Freethorpe, the Broads National Park

Fig. 2.2.8 High's Mill, 2014, Halvergate Marshes, the Broads National Park

Fig. 2.2.9 Cadges Mill (foreground) and Berney Arms Mill – not all at-risk mills can be restored as part of the project; 2019, Reedham, the Broads National Park

Chapter Three
DESIGN

INTRODUCTION

For many landscape architects the design stage is where the fun begins. We are designers, and design is the focus of our training and likely to be what drew us to the profession. In reality design work might only be a small element of our workload, although we will probably have been developing a design in our heads from the moment we visit the site. True designers are constantly redesigning their environment in their mind's eye, driven by an innate dissatisfaction with poor design.

During our training we are taught that the design process is linear, with distinct stages such as survey and assessment, and clear transitions between each stage. That may be how it is articulated to the client but for most designers it is a more scattered, subconscious process with ideas triggered by a diverse range of factors. The more experienced we are the easier the process appears. A line on a page may look random but it is based on experience and a full understanding of the implications. Good design in landscape architecture can mean that the scheme recedes rather than dominates, meaning that it doesn't gain the attention of an iconic building or a statement bridge.

As we move into the design stage the brief will be finalised, all specialist sub-contractors appointed (in some cases we might be that specialist sub-contractor), consultation completed and costs agreed.

Fig 3.0 (Pages 68 and 69) Almhöjden, Skogskyrkogården (woodland cemetery) designed by Gunnar Asplund and Sigurd Lewerentz' built 1915 to 1940 and designated as a UNESCO World Heritage Site in 1994, 2014, Stockholm

STAGE 2 – CONCEPT DESIGN

- Develop landscape and ecology strategy
- Agree masterplan/concept, circulation and management strategy
- Develop concept design
- Monitor progress of concept design
- Prepare and issue final project brief
- Project team meeting
- Review risk assessments with project team
- Review handover strategy with project team
- Review and update project execution plan
- Review project programme and agree any changes with project team
- Comment on stage design programme
- Comment on cost information
- Prepare sustainability strategy
- Develop green infrastructure strategy
- Prepare management plan
- Prepare stage design programme
- Monitor and review progress and performance of design team
- Liaise with planning authorities
- Stakeholder engagement
- Undertake third-party consultations
- Assist lead designer with preparation of stage design programme
- Provide information for preparation of cost information
- Prepare construction strategy
- Develop health and safety strategy
- Formal sustainability pre-assessment and identification of key areas of design focus
- Initial energy assessment
- Environmental impact check
- Evaluate landscape effects and visual impact effects of development
- Climate change check
- Develop landscape and ecology strategy
- Design team meeting
- Carry out design/technical review
- Prepare masterplan
- Produce implementation strategy for masterplan
- Develop outline/performance specification
- Exchange concept design

Table 3.1 The elements of this workstage[1]

STAGE 3 – DEVELOPED DESIGN

- Agree spatial arrangement, materials palette and planting character
- Monitor progress of developing design
- Review handover strategy and risk assessments with project team
- Review and update project execution plan
- Project team meeting
- Review project programme and agree any changes with project team
- Comment on stage design programme and cost information
- Coordinate and comment on design proposals and project strategies as they progress
- Update sustainability strategy and maintenance and operational strategy
- Prepare stage design programme in conjunction with other design team members
- Prepare developed landscape design
- Develop detailed specification
- Liaise with planning authorities
- Prepare landscape management plan
- Develop planting strategy
- Submit planning application
- Undertake third-party consultations
- Provide information for preparation of cost information and project strategies
- Interim energy assessment
- Resource and waste minimisation design review
- Stakeholder engagement
- Design team meeting
- Carry out design/technical review
- Exchange developed design
- Prepare mitigation proposals informed by the LVIA

STAGE 4 – TECHNICAL DESIGN

- Complete hard and soft detailing
- Prepare tender documents for construction and maintenance
- Monitor progress of developing design
- Review updated handover strategy, project strategies and risk assessments with project team
- Project team meeting
- Review and update project execution plan
- Comment on stage design programme
- Manage change control process
- Monitor and review progress and performance of project team
- Prepare stage design programme in conjunction with other design team members
- Monitor and review progress and performance of design team
- Liaise with specialist sub-contractors
- Prepare landscape technical design
- Submit Building Regulations Submission (Building Warrant in Scotland)
- Undertake third-party consultations
- Assist lead designer with preparation of stage design programme
- Update construction strategy
- Assist in preparation of landscape contract, agree with contractor and arrange completion
- Check sustainability assessment
- Check energy assessment
- Check user guide
- Check outstanding design stage information
- Check monitoring technologies
- Check review of changes
- Check sustainability compliance
- Co-ordinate design proposals and project strategies as they progress
- Design team meeting
- Carry out design/technical review
- Develop technical specification
- Exchange technical design

THE LANDSCAPE ARCHITECT

A garden is a complex of aesthetic
and plastic intentions; and the plant is,
to a landscape artist, not only a plant
– rare, unusual, ordinary or doomed
to disappearance – but it is also a color,
a shape, a volume or an arabesque
in itself.'

Roberto Burle Marx

As mentioned in the previous chapter, we design in part with living material. Our work includes the fourth dimension of time, a concept that can be difficult to grasp. We must design a scheme that works when the site opens, but also works as the living components grow and develop. The timescales for landscape architects are longer than those of other construction professionals, with even the most short-lived tree species often living for a century and the longest growing for millennia.

Design stages

In his excellent 2017 book *Slow Growth: On the Art of Landscape Architecture* landscape architect Hal Moggridge discusses the order his practice uses for making design decisions.[2] Hal worked with Brenda Colvin as principal of Colvin & Moggridge, the oldest surviving British landscape practice, and is a world-renowned expert on landscape design, with projects including Blenheim Palace and the Botanic Garden of Wales.

Hal's stages show that the setting is assessed first and then the building sited – in this method buildings sit in rather than dictate the landscape.

Design is rarely this tidy, ideas often evolving in a less structured way, but having a framework of stages to work through gives the project team a systematic process to meet the client's brief and ensures each issue is resolved in a sequence that doesn't preclude the next stage.

Projects where the issues of routes between destinations are not resolved at concept stage, perhaps where a landscape architect has not yet been appointed, can end up with awkward layouts that in turn cause unnecessary issues. These might be areas of a site with no real purpose, or features that should be screened, such as car parks, being placed in the most prominent location.

Hal's sequence deals with the fundamental structure of the scheme first, ensuring that it works at the most basic level, and then adds layers of finesse and refinement once the structure is in place. It deals with the issue of movement routes around the site, one of the central requirements of most landscape schemes, and looks at how the experience of the user moving through the site can be enhanced.

'A landscape plan is frequently
a plan of movement routes.'

Hal Moggridge,
Slow Growth: On the Art of Landscape Architecture

1. Key elements are located, including the points of entry onto the site, existing features to be retained, activities needing level land such as playing fields, principal waters; then buildings.
2. Main routes connecting these destinations are defined.
3. The principle open space system designed, taking into account land uses, views, etc.
4. The main tree masses are located, both to define the spatial composition, including buildings, and to hide intrusive elements and irrelevant views.
5. Secondary destinations are located, whether secondary buildings or functional spaces.
6. Subsidiary routes are defined.
7. Detailed enrichment is designed, such as planting, streams, cascades, playgrounds, notices, seats, etc.

Hal Moggridge, *Slow Growth: On the Art of Landscape Architecture*, London, Unicorn, 2017

Presenting our design

Our designs are useless unless they can be interpreted by others. Design education can risk focusing on the visual representation of the design, such as the quality of plans and models, rather than the quality of the design as it will appear on site. With no final scheme on site to review drawings form the basis of assessment.

This gives an over-emphasis on the plans over the final design – our plans should be the least information needed to get our ideas across. They are rarely required to be work of art in themselves. They are a visual description of what is in our head.

Our job is to take our client's needs, gain a full understanding and then develop a design that meets those needs. We then relay that design back to the client, gain their approval, and then present the design in a way that can be reviewed and built by others.

The difficulty is getting the idea from our head into the head of our client and the rest of the project team. That idea is then passed on, potentially down a chain of people who we may never meet, such as estimators or suppliers. Quite often we must convey that idea to another person when we are not physically present to explain, in the same way that writers convey ideas in their work. In his excellent book *On Writing* the author Stephen King describes this as a form of telepathy.[3] He illustrates that idea by describing a rabbit in a cage, with a number drawn on its back. Both the writer and the reader have an idea of the rabbit but their idea of the size of the cage, or the colour of the number will be different. In fiction that difference is what exercises our imagination but in design work that difference needs to be as small as possible.

Using clear plans and careful specifications we can get close to what we envisaged but unless we undertake all the work on site ourselves or spend a disproportionate amount of time supervising the work it is unlikely that it will be a perfect match. With a talented project team it may be better than we imagined, and hopefully the difference is so minor as to be acceptable.

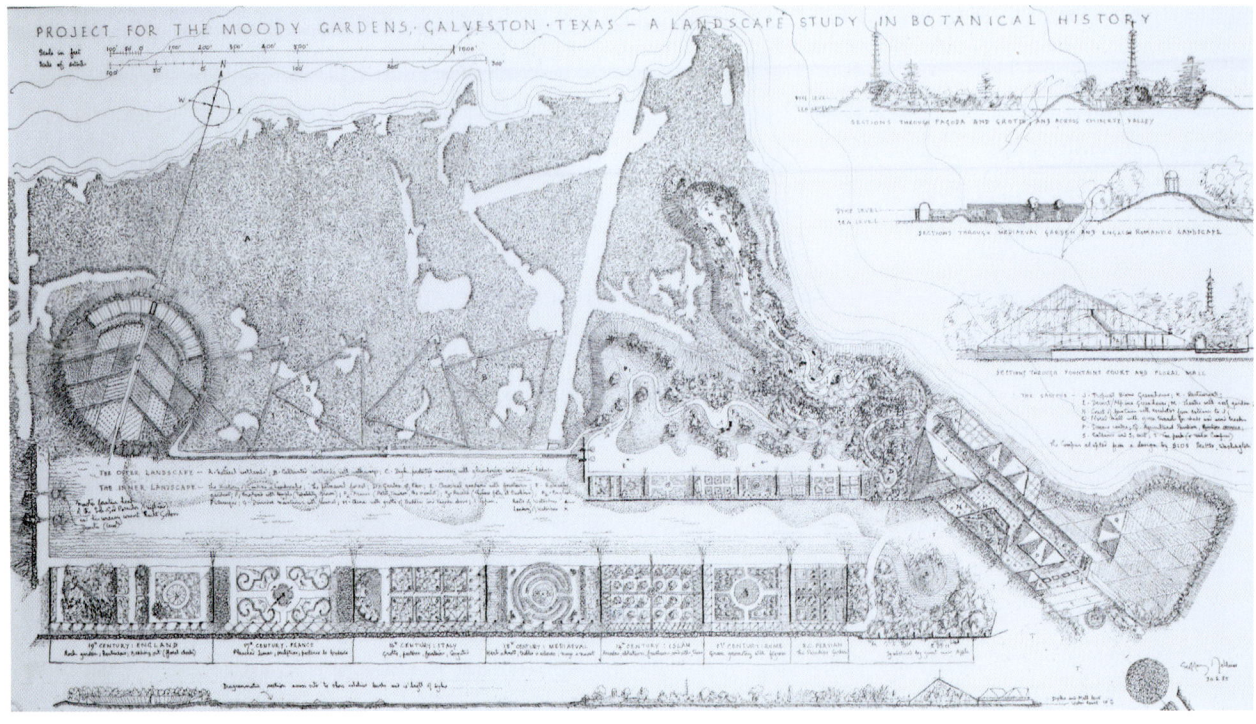

Fig 3.1 Design for the Moody Gardens; Sir Geoffrey Jellicoe, 1985, Galveston, Texas

Even if we use 3D drafting tools to design our work, it is likely that in most situations that design will end up as a 2D representation, either on paper or a digital equivalent.

The level of detail required to allow a full understanding of our idea needs to be balanced with the amount of information shown. A good plan is clear, easy to understand and allows minimal opportunities for ambiguity. Our training can encourage us to believe that plans are a creative outlet rather than a tool in the creative process.

Creators of plans demonstrate their own agenda by what they include and what they exclude. An unverified mapping myth suggests that early maps produced by the UK Ordnance Survey for the military used two different line types for walls and other boundaries to show whether they could be used as cover to hide behind. Plans drawn for children would show safe places to cross a road, play areas or bus stops, whereas most mapping focuses on more adult concerns such as road networks, golf courses or pubs.

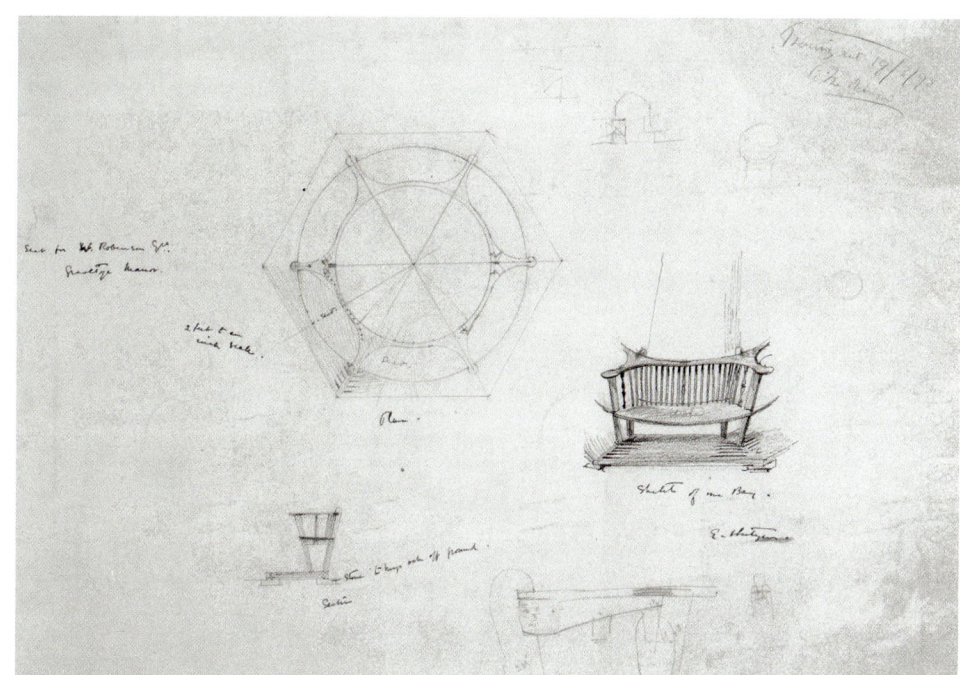

Fig 3.2 Design for a wooden circular seat at Gravetye Manor; Sir Edwin Lutyens, 1898, West Hoathly, Sussex

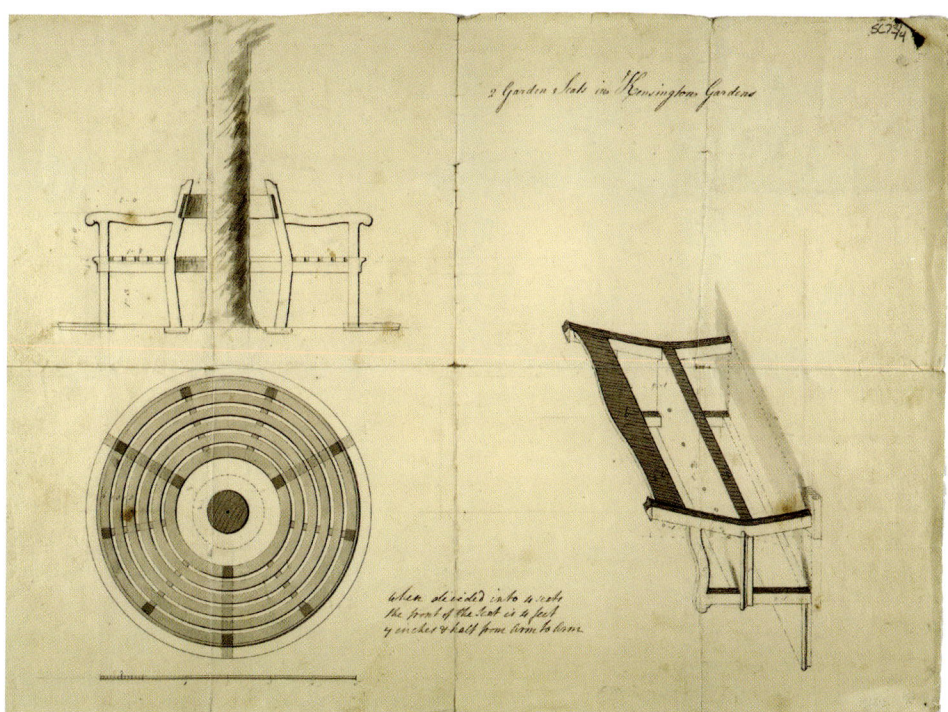

Fig 3.3 Designs for gardens seats, Kensington Palace Gardens – plan, section and perspective; John Vardy, 1750, London

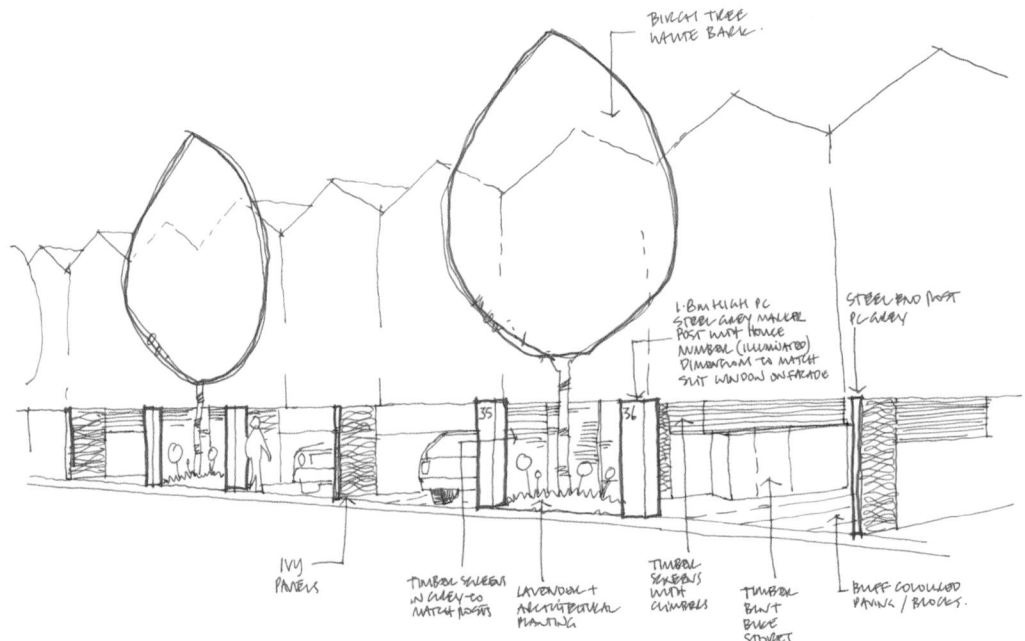

BIRCH TREE
WHITE BARK.

1·8m HIGH PC
STEEL GREY MALEAR
POST WITH HOUSE
NUMBER (ILLUMINATED)
DIMENSION TO MATCH
SLIT WINDOW ON FACADE

STEEL END POST
PC GREY

35

36

IVY
PANELS

TIMBER SCREEN
IN GREY TO
MATCH POSTS

LAVENDER +
ARCHITECTURAL
PLANTING

TIMBER
SCREENS
WITH CLIMBERS

TIMBER
PLANT
BOX
STOREY

BUFF COLOURED
PAVING / BLOCKS.

This bias can come across in the surveys or base plans we receive, such as a survey that shows existing vegetation as amorphous blobs with ill-defined boundaries, or trees with identical, symmetrical canopies. Another example is the 'frozen pea plan', a landscape concept plan that shows trees as tiny green circles scattered across a site with no reference to final height or root zone. This minimisation of the potential landscape scheme can be a clue as to how the landscape is valued by the plan creator.

Drawing conventions and standards, such as those created by the British Standards Institution (BSI) and the International Organization for Standardization (ISO) (BS EN ISO 11091:1999 Construction Drawings. Landscape drawing practice) can help reduce ambiguity and ensure plans are clear and uncluttered.[4] Standardisation, such as the conventions introduced in Building Information Modelling (BIM), is seen by some as restricting creativity but if it reduces errors in the interpretation and execution of our work it should be seen as a positive step.

When deciding what to include in a plan we need to identify the ultimate use and include only what is needed to achieve that aim. A talented draftsperson can convey an idea with just a few strokes.

The mechanics of design

The process of producing plans, 3D models or reports should not get in the way of the design process. When computer-aided design (CAD) started to be used in the construction sector the death of design was predicted, but the reality is that we can now create buildings and structures that would have been unimaginable, such as the parametric structures created by architects Zaha Hadid and Frank Gehry. The ease of amending plans means we can quickly test different design options. Digital plans can easily be shared, with different team members collaborating in real time on even the most complex 3D models.

Our aim should be to improve our productivity, reducing the time spent on laborious, low-skilled tasks such as manually creating planting or materials schedules, so we have more time for the creative aspects of our work. By improving our efficiency we should improve our productivity and hopefully make more profit by reducing the time taken to produce plans and specifications. Regularly reviewing the workflow for each type of document you produce is a valuable use of time, exploring whether innovations could reduce the production time.

Fig 3.4 (Page 76) Design for a house for an art lover; Charles Rennie Mackintosh, 1901, Scotland

Fig 3.5 (Page 76) Design for a housing development; David Jarvis Associates, 2014, Swindon

WORKFLOW FOR LANDSCAPE AND VISUAL IMPACT ASSESSMENT PLANS

Landscape architectural practice David Jarvis Associates identified that using geographic information systems (GIS) and freely available open data could dramatically reduce the workflow for the creation of the plans needed for Landscape and Visual Impact Assessments (LVIAs). The practice, founded by past President of the Landscape Institute David Jarvis, has been involved in the evolution of LVIAs, contributing to the most recent edition of the book *Guidelines for Landscape and Visual Impact Assessment*.[5]

Their original workflow was to take the source information, such as the boundaries of designations, from PDFs and paper reports. These would be brought into drawing software and the lines redrawn in the new format.

As the source data they used was only provided as a single layer plan, the digital equivalent of a paper plan, they couldn't separate out the information they needed for each document.

The UK government policy of making public data freely available now means that much of the information needed for LVIA work is available as digital data sets. Open data, from the Land Registry, the Environment Agency and other public bodies, is now online and can be downloaded for free.

GIS software allows the user to create maps that include data and are made up of points, lines or polygons. By creating template base plans in GIS software and then connecting these to the open data, accurate plans can be created in a short space of time. The data set usually has numerous layers which can be turned on or off, so the relevant information can be included and any superfluous detail turned off. For information that is regularly updated, such as the land ownership data provided by the Land Registry, the source data can be accessed via a live web feed to ensure that the information is up to date.

By reviewing their workflow David Jarvis Associates reduced the time taken to produce a set of standard plans for an LVIA from a few days to just a few hours. As the original source information is used, the risk of errors when copying is removed and the information is easy to update.

Open data and analysis

Before GIS software became available to the landscape architect the only way to compare different data sets to analyse a site was through sieve analysis. This usually involved layers of tracing paper over a base plan, and careful comparison of the layers would reveal areas of concentration or deficiency of the factor being assessed.

GIS now means that complex data sets can be put together, a query created to interrogate the data and insight gained that would have been previously outside the scope of a landscape practice. Queries are similar to formulas in spreadsheets – the user selects a list of criteria and the software displays the features or records that meet that criteria.

This level and speed of analysis means we can quickly assess potential routes, site assets, levels of engagement or areas with a deficiency of access to open space. If we have 3D data we can use GIS software to assess Zones of Theoretical Influence (ZTIs) and watersheds.

The cost of this analysis would have been prohibitive to many landscape practices, as both the software and the data sets were expensive. In 2009 the free, open-source program Quantum GIS, or QGIS, was released. Previously seen as only relevant to those with a high level of technical knowledge, and potentially unreliable, open-source software is now widely used, including by public-sector organisations.

Access to open data and continuing improvements in GIS systems mean that we can use technology to analyse the landscape in ways we could never have conceived, in a short timescale and at minimal cost. It does take time to learn to accurately use the tools but this is offset by the insight we can gain and the time we save compared to previous techniques.

All openly available data should be assessed for credibility – the computer science maxim of 'garbage in, garbage out' applies to open-source data just as much as it does to any other source. The UK Landscape Institute has a useful list of data sources for landscape architects.[6]

There may be insights we have yet to discover, that can only be unearthed by overlaying divergent and seemingly unconnected data sets. If we use technology effectively we may discover correlations, be able to question assumptions and demonstrate the value of the work that we do.

Inspiration

Landscape architecture is an art, and it is entirely valid for us to take inspiration from a wide range of sources. We don't need to share these with the project team or the client, unless they are part of a wider

MAGIC.DEFRA.GOV – INTERACTIVE ONLINE GIS

An early example of using open data with GIS is the UK government website www.magic.defra.gov. Set up in 2002 the site uses interactive mapping to allow users to find information about a location, drawn from up-to-date government sources. It is simple to use, doesn't require specialist software and runs in a standard web browser.

Users can set criteria for their search by selecting which data sets to include and setting the boundary or radius of their search area. The numerous data sets include administrative boundaries, habitat designations, agricultural land classification, planning designations, public rights of way and scheduled ancient monuments.

Users can save a link to the map they have created or export their search as a map in a number of formats. The data sets are also accessible from the site, allowing users to easily find the underlying data for use in their own GIS maps.

concept that the whole team will work to, but they can add depth to our work.

Geoffrey Jellicoe drew on a wide range of ideas for inspiration, such as Jungian principles and modern art. One of his most famous schemes, the John F Kennedy memorial site at Runnymede, an acre of English ground given to the United States in honour of the assassinated President, explored the concept of the visible and invisible world. To Jellicoe the invisible world included allegory, symbolism and the subconscious. The design includes a stepped woodland path made of granite setts, with 50 steps each representing the US states, along with religious and literary references.

'... if you are proposing to create a landscape allegory, you must not, by explaining it, bring it within reach of the intellect (as I am recklessly doing tonight) for then it could easily be so dissected as to become trite.'

Geoffrey Jellicoe Lecture[7]

The use of symbolism is commonly accepted in designs linked to memorial and remembrance. The design for the UNESCO World Heritage listed Skogskyrkogården (woodland cemetery) in Stockholm, designed by Sigurd Lewerentz and Gunnar Asplund' started with the experience of visitors – the concept of mourning and the feelings surrounding it. The cemetery is simple and beautiful, with specific places for reflection such as the Almhöjden (elm heights) meditation grove.

Fig 3.6 John F Kennedy Memorial; Sir Geoffrey Jellicoe, 1965, Runnymede, Surrey

Fig 3.7 (Page 81) Steps to Almhöjden, Skogskyrkogården in Stockholm, a UNESCO World Heritage site. The rise of each step decreases as the path climbs, slowing the visitor down, stopping them becoming tired and calming them in preparation for the place of mediation in the grove at the summit; Sigurd Lewerenz and Gunnar Asplund 1917–1940, Stockholm

Fig 3.8 Planting scheme for butterflies and other pollinators – wildlife-friendly planting is one way landscape architects can add value to a project; Rosendals Trädgård 2014, Stockholm

Sources of inspiration

Making sure there is space in our working day to get away from our desk and to find inspiration can be a challenge but it is important to recognise that what can seem like unproductive time is part of the design process. The 'white space' between formal activities can be the time our subconscious uses to process ideas and find solutions.

Designing is very close to daydreaming. Research by Dr Sandi Mann and Rebekah Cadman at the University of Central Lancashire in the UK found that rather than being a wholly negative state short periods of boredom can trigger greater levels of creativity.[8] We might not be aware how valuable this time is – the short walk or the

mundane task – until we cut them from our routine and our creativity drops, and problems become harder to solve.

'Boredom is not an end product, is comparatively rather an early stage in life and art. You've got to go by or past or through boredom, as through a filter, before the clear product emerges'
F Scott Fitzgerald, *The Crack-up*

Having a diverse range of topics and learning about issues outside our immediate profession can help creativity. Working methods change and other sectors can teach us new methods and give us ideas that can spark innovation. Could the space sector teach us how to handle vast quantities of data? Could the digital animation sector teach us how to collaborate effectively? Only ever having existed in digital format and requiring high turnover, they have no paper legacy to limit their working methods.

When technology is a distraction it may not aid our creativity, but the ability to access information on almost any topic on a device no larger than our hand means we have at our immediate disposal a wealth of inspiration our predecessors would envy. Podcasts, blog posts, social media updates and credible websites are all potential sources of inspiration, with the element of chance that can help us find a solution to unblock a problem.

We rarely have the luxury of time to develop our ideas, but allowing as much time as possible for the design process means that all options can be explored and thought through, and also allows more time for the serendipity, such as a relevant online article or a chance conversation, that can sometimes provide a solution to a difficult design issue.

Adding value

As the brief is finalised we have our last opportunity to make sure that the client understands how we can contribute to the project. If they don't know that we can add habitat value, sequester carbon or manage storm water they can't appoint us to undertake that work. Making clear the added value the landscape architect could bring to the project ensures that opportunities aren't wasted. The client may not be able to fully utilise our skills on this project but will then at least have a better understanding of our potential role for any future work.

We also need to make sure that the client and the project team know when we add value to a project. If we include additional features within the project budget we should explain this benefit, so the value of our work is recognised. The value of good landscape design can often be hidden so unless we articulate that benefit we can't criticise others for not recognising it.

And, just as importantly, we need to recognise that value ourselves – it can be easy to overlook the value we bring to a project. Some of the subtle refinements we make to a design can add substantial value, such as reducing maintenance costs; refinements we make almost subconsciously based on our experience and expertise. The value we bring to a project should be recognised, both socially and financially.

Maintaining our role

As the project progresses it is important to ensure that the design roles have not become blurred, either with aspects of work being undertaken by others, or by extra work added into our remit with no recognition. This can be difficult to identify if there is a complex hierarchy within the project team, as it can prevent us from having an overview of how the work is being distributed.

Sticking to the brief can be a challenge. New issues can arise that mean the original concepts no longer apply – in this situation the brief needs to be revised rather than ignored, so roles are still clear.

Specialist areas of design

As part of the design process the landscape architect may be able to provide more specialist design skills that the client or members of the project team are not aware of.

Some examples include:

Design for people with dementia – The neurological changes experienced by people with dementia can affect how they perceive the world around them, including the surfaces they walk on.[9] Sharp contrasts between light and shade on a paved surface can appear as if they are made from different materials. Pictorial signage can also become confusing, such as male and female symbols for toilets rather than a toilet symbol.[10] Small changes to a design can make a space easier to navigate and more inviting for those with dementia.

Public safety – Changes in terrorist activity over the last decade have increased the focus on the safety of public spaces. Vehicle barriers are now commonplace in shopping areas as well as around important government buildings. Landscape architects can help ensure that the design of safe open space is in keeping with the character of the landscape, and that any defensive structures are integrated into the landscape scheme. Case Study 4.1 covers the issue in more depth.

Landscape resilience – Through our work we have the opportunity to make the landscape more resilient to climate change, natural disaster or future changes in use, such as population growth.

Designing for the future

Given the materials we use we will be thinking about the long-term use of the site, and how elements might look in future centuries. We can't predict the future but we should give some thought to how future changes might impact on the schemes we design. As mentioned earlier we may need to look at making our schemes more resilient to higher temperatures, or heavier rainfall, due to the impact of climate change.

Matching plants to existing site conditions whilst accommodating future conditions can be a challenge. It may be that an area that is at the northern or southerly limit of the range for a species may be outside that viable range in the future. One good example is fruit varieties that were grown to suit the conditions in a location, such as the Blenheim Orange apple. Varieties thrived in the conditions found in one area, such as rainfall or late frost. Climate change means that the climatic conditions in that location might no longer apply, with a change in growing days or the date of the first or last frost. If the plant has a very narrow tolerance it may no longer grow it in that location in 100 years' time.

Fig 3.9 Light and shade can be difficult for people with dementia to negotiate; Parc Cefn Onn run by Cardiff City Council, 2019, Cardiff

This highlights the importance of knowing the requirements of a plant, considering the habitat it comes from and the conditions it can withstand.

Seeing a plant growing in its optimum habitat can be a valuable education, with specimens often looking quite different from the plant on site that you previously felt was thriving. Trying to identify a tree in a more northerly climate and realising that it is a species you know well but in a cooler, drier climate grows much larger with more foliage, or seeing a shrub that is small and compact in your climate grow large and bushy in an area of high rainfall, is a useful reminder that we are often growing plants in less than optimal conditions.

The opposite situation can also occur – plants that have a balanced role within an ecosystem in their natural habitat can become invasive and dominant when introduced into a new habitat which favours them. With limited challenges to growth these introductions can damage important native habitats. In the UK species that were introduced into gardens and became invasive include Japanese knotweed and Himalayan balsam.

With the number of variables impacting on plant health, such as competition, soil type, health and depth, temperature range, rainfall, wind speeds, competition and air quality, it isn't easy to work out which factors are affecting growth but it is still worth reviewing, especially with the anticipated impacts of climate change.

Another change that is expected to have a significant impact on the streetscape is the introduction of autonomous vehicles – will we keep our vehicles right

Fig 3.10 Portland stone planters and bollards to prevent vehicle attack; Broadcasting House, 2018, London

outside our homes, or will we store them out of sight and call them up when needed? Will domestic garages become obsolete? Will there be an increase in vehicle movements as empty vehicles travel to pick up their owners? Or will there be a decrease in vehicles with smaller, flexible and more frequent public transport made possible by the saving in driver costs?

The move to autonomous vehicles could trigger a change in housing design, in the same way that the rise in personal car ownership in the 20th century led to the design of the suburbs.

We need to watch trends and try to accommodate these as much as possible, without being distracted by issues that may come to nothing.

Chapter Four
THE DESIGN
TEAM

THE CLIENT

During the design stage the client must trust that the concepts they expressed and the parameters they set have been understood by the design team, and that they are adhered to as the design develops. The client is usually taking the greatest risk, and relies on the professionals they appoint to mitigate that risk as much as possible. The design team also help the client maximise the potential of a project, perhaps by identifying overlooked opportunities or by using new techniques.

Working with the client

Being the client can be difficult. It can be easy to forget that the client may be struggling to fulfil this role – if the project is not in their area of expertise it can be hard for them to identify what is wrong and articulate their concerns. If you consider situations outside of your work life where you act as a client, such as having a car repaired or calling an emergency plumber, you can feel almost held to ransom by your lack of knowledge or the urgency of your need.

Even in our working lives something seemingly as simple as redesigning a website can be hard to manage. You have an idea of what you want but don't know how expensive that is to achieve, or the practicalities of managing it day to day. You are dependent on the designer to guide you and this requires a high level of trust. If a relatively inexpensive process such as

Fig 4.0 (Pages 86 and 87) Merton Borders, June 2014, University of Oxford Botanic Gardens, Oxford

redesigning a website can feel fraught, imagine how being a client on a large construction project might feel.

The client–designer relationship doesn't need to be close but both parties need to trust each other for the project to succeed. To take what is usually a verbal description of an idea and from that create plans for a three-dimensional scheme requires the client to have confidence in their design team.

Our clients need to understand that we need to hear their concerns as early as possible, and that changes at the design stage are far better than changes during construction. To ensure that they feel they can voice their concerns we need to create a system where they don't feel that changes or comments are criticisms that will be taken personally.

The client's role in the design process

The client is central to the design process, as the design has been instigated to meet their need. The RIBA Digital Plan of Work overview doesn't mention the client in any of the design stages. In fact the client is only mentioned in 0 – Strategic Definition and 7 – In Use, which seems an oversight given that without a client there is no project.

There are many tasks that the client fulfils in the design process once they have set the brief but the main roles are:
- supply of information, including information about potential health and safety issues[1]
- budget
- timescale
- decision making and level of development – what can we decide and what does the client want to

decide? Balance of our expertise (why appointed) and making sure the client is happy with the result
- ensuring approvals and consents are agreed
- setting priorities
- scrutiny – is the design on track? Is it what they want? Is it meeting the standards they set out?

Supply of information

Assumptions are easily made – we can assume the client owns or has the right to develop the whole site, that they have the authority to approve the project or that this is the first time they have embarked on this design. Only by asking fundamental questions that we might assume we know the answer to, or that the client might mention, can we uncover the full situation. Discovering that the client owns only part of the site when you are preparing planning drawings is not ideal.

Encouraging the client to provide all the information they have, however irrelevant it seems to them, can reduce the risk of surprises later in the project. There may be issues that the client can't or won't tell us, such as how the project fits with a wider strategy or the long-term aspirations for a site, but we need to find out as much detail as early as possible to uncover any issues that might be a constraint on the design. Building up a checklist of questions can help draw out issues.

Budget

The budget can be one of the largest sticking points within a project. Some clients are reluctant to share how much they intend to spend, making it difficult for the designers to know how to fulfil the design.

INFORMATION LIST

Possible issues include:
- land ownership
- site access and covenants
- rights of way and permitted access
- underground services
- soil contamination
- previous land use
- previous ownership
- project history, including options that have been disregarded
- business case and agreement for the project
- project approval
- planning history, including any pre-application discussion with the planning authority
- long-term site plans or aspirations
- the Do Nothing option

For all construction projects the potential range in cost of a design can be substantial, depending on the time allowed for construction or the quality of materials used. With landscape architecture that range is just as great, as a single tree can cost a small amount of money if supplied as a young whip and planted into open ground, or a significant amount supplied as a semi-mature tree with complex tree pits and guying systems. Even clients who say they have no idea about budget will have some idea if pressed, even if it is a bracket that they need to stay within. A mismatch of expectations, with the client expecting to spend a fraction of what the landscape architect knows is needed to fulfil their needs, is not the basis of a successful project.

There is a culture within some sections of construction of allowing clients to believe that they can achieve their ambitions with an inadequate budget. This raises the client's expectations, which are then almost inevitably dashed as the design is value engineered to fit within the final budget. This is a difficult situation, as telling clients that they can't achieve their aim within their budget is not a good way to win bids. However, it is one of the major weaknesses within our sector – Mark Farmer, author of the Farmer Review mentioned in Chapter 2, stated that often the person who wins a construction contract is the one prepared to submit an unrealistic price.

The budget may fluctuate as the project progresses, perhaps with the landscape budget being cut as building costs increase or with more money allocated as savings are found elsewhere. It is important to agree how the budget is controlled and the project team updated. It can also be useful to decide with the client what their priorities are if the budget is reduced – is hitting the timescale the most important factor for them, or will they allow more time and focus on quality?

For landscape projects the client needs to decide their budget for maintenance of the site, including the establishment period. A lower cost scheme that is well maintained is always better than a more lavish scheme that receives no care. A set maintenance period may be a requirement of any planning permission but this is usually a minimal period compared to the potential life of the scheme. A good maintenance plan should show the potential life span of the plants and materials used,

and guidance on replacement and succession planting. For tree planting that succession planting may be in 200-year timescales and outside the control of many clients, but it is important to make the point that the materials we use are living materials with a finite life span.

Embarking on a project with so many variables and risks does make predicting the final budget a challenge, but advice on realistic contingency percentages and careful risk mitigation can help to reduce the uncertainty.

Timescale

Like budget, timescale can be difficult to agree. There can be a risk of working to an arbitrary timescale, decided early on in the project with limited scrutiny, when a longer timescale would have been more suitable and potentially saved costs, such as better prices from contractors, allowing smaller plant stock or seeding rather than turfing a site.

The project team need to agree with the client what is critical and what is flexible. Explaining seasonal issues that can impact on timescale, such as restrictions on site clearance due to bird nesting, the bare-root planting season or the times of year habitat surveys can be undertaken, can help the client agree a realistic timescale. Discussing the risks of the planning process, and giving a realistic assessment of the time this may take, can help manage timescale expectations.

Decision making

The client's role changes from explaining their needs to deciding on options that could fulfil that need. Often the two roles will run in parallel, with the client reviewing their needs as new options or constraints arise. Despite taking on a professional design team the client is still likely to need to make numerous decisions, each with cost and quality implications, as the design process develops.

The client needs to be encouraged to voice concerns as early as possible. This can be a difficult habit to develop, as the client may not be able to articulate the problem if the issue is outside their area of expertise. Again, think back to a situation where you have been a client and there was an imbalance in knowledge – you might know your car isn't running as well as it should but with no specific fault you don't know where to direct investigations, and you are worried about expensive testing that might not identify the fault. Our clients can be in a similar position, but with far higher values at stake. We need to develop a relationship where comments and changes are not seen as opening up the risk of greater costs, or offending the designer.

Working with the client to forewarn them of the number and nature of decisions needed, and working out an efficient process for managing these decisions, can be time well spent. Agreeing what level of decision the client wants to take and how much they want to delegate can reduce the number of queries for the client – do they want to approve every detail or are they happy to approve concepts? Do they have particular views on plant selection? Understanding the client's capacity and working patterns, and how they like to

IF THIS THEN THAT

If This Then That (IFTTT) is a free online automation platform that allows users to connect hundreds of different apps and devices to create combinations that run automatically.[2] Primarily aimed at domestic users it also connects to project management and collaboration tools such as Basecamp, Slack and Trello, as well as other online tools including Google Docs and Dropbox. Users create applets based around certain conditions. An applet might automatically back up the photographs on your phone, or send you an email when the weather forecast is good, so you can plan a landscape and visual impact assessment.

Using IFTTT it is possible to create an easy-to-use, automated system that allows decisions to be recorded. Applets could include:
- if I receive an SMS with the tag decision, save it to a Google spreadsheet
- if I receive an email to my Gmail account from x with the word 'decision' in the subject line, save it to Dropbox
- if I receive an email labelled 'decisions' post it in the decisions channel in Slack.

IFTTT can also be used for staff on site to record that actions are complete – at the end of the day they could send an SMS to record the work completed that day or save photographs for a snagging list to Trello.

IFTTT applets can even connect Internet-enabled devices, such as weather stations and irrigation systems, allowing localised, automated irrigation.

communicate, can help tailor a system of approval that works for all parties.

Agreeing communication methods with our client at an early stage can help encourage them to mention concerns as they arise. It could be via an online project management system, a shared spreadsheet, a weekly email, a messaging app or even a text message. The system needs to be appropriate for the whole team and to create a record that can be referred back to if decisions are ever disputed. A balance needs to be found between making demands on the client's time and receiving enough feedback to create a scheme they are happy with.

If we are part of a complex sub-consultancy communication can be even harder to manage as we may be reliant on others to act on behalf of the client. Creating a relationship where the client feels they can speak up can be difficult but we need to find a way to communicate effectively and understand their needs.

Like all other activities it is a good idea to regularly review workflows, looking for ways that routine tasks can be simplified or automated. As technology changes, services that were once only available to large companies with their own software developers become mass market, and are sometimes even free to use.

Priorities

The client will set the priorities for the project. These should never be assumed – we might assume that a public-sector client wants to deliver the lowest-cost scheme possible but they may want to look at whole-life costs, including maintenance, and spend more upfront to reduce their long-term liabilities. Other clients may

have a timescale that is immovable and for them time is likely to take highest priority. Unless we discuss the client's priorities we won't know what they are.

Information and updates

Earlier in this chapter we looked at how decisions can be tracked. We also need to work out processes for how information is managed, including updates. Managing data is one of the main functions of BIM, using a Common Data Environment (CDE) to centrally store and control all project documents.[3] Problems caused by the wrong version of plans being used are embarrassingly common in the construction sector. If we want to improve the image of our sector and also deliver the scheme that the client has agreed, we need to manage basic processes such as document issue.

Managing drawing issue can be as simple as a basic spreadsheet, duplicating the function of the paper templates used before we issued drawings electronically. But with a digital process it is possible to automate the control of documents and allow that information to be shared within the project team.

Scrutiny

Having instigated the project and set the parameters the client has to scrutinise whether progress is being made and if it is to the standard they expect. A good brief at the outset makes this process easier for all parties, as the standards expected will have

Fig 4.1 (Page 93) Succession tree planting, with new trees protected by timber tree guards, 2014, Stowe Landscape Gardens, Buckinghamshire

been defined. The client needs regular and accurate information to allow them to assess progress and approve work. The client may be being scrutinised by their managers or a board, so the information provided needs to be tailored to provide the correct level of detail for all those involved in monitoring the project.

The client–designer relationship

There is limited research on the client/designer relationship and how to cultivate it effectively. For such a central part of our work it deserves further investigation. The RIBA report 'What Clients think of Architects – Feedback from the 'Working with Architects' Client Survey 2016', which consulted almost 1,000 clients found that:

- 76% of private domestic clients, 73% of commercial clients and 51% of contractors were 'very' or 'fairly' pleased with their project, overall
- 61% of private domestic clients, 56% of commercial clients and 30% of contractors were 'very' or 'fairly' satisfied with their architect's process management performance.[4]

However, at the time of writing no guidance or new policy has been produced to help remedy the situation.

A greater understanding of this relationship, central to most projects, could reduce misunderstandings, help create better project briefs, improve client satisfaction and potentially prevent disputes.

Our training, including our professional qualifications, tends to focus on practical issues, such as legislation, health and safety or contract processes. As we develop our careers it is worth looking at other skills, such as project management and communication techniques, that could help us work more effectively with our clients.

Sharing the design with the client

A landscape design, in three dimensions and with elements that will change over time, can be hard to express in plan form. Complex cut and fill plans, showing existing and proposed contours, can be difficult to interpret. As mentioned earlier, the function of a plan is to take an idea from one person's mind, summarise it, and then represent it in a way that others will interpret in as close a way as possible to the originator's idea.

Deciding how to show the design, in what level of detail and at which stage, can be a difficult decision. We need to balance time taken to illustrate an idea which is only used to make one decision against the timescale for the project and the fees allocated, and the need to ensure the client understands what is being proposed. There may be versions of the plan for the client, with images to show planting styles or styles of street furniture, and a more functional plan that is used for planning discussions or team meetings.

Technology can help greatly when creating different versions of the same design. In drafting software the underlying design data can be the same with separate plan views created for each type of use, tailored to the specific audience.

The same drafting software can also represent the design in a proxy for 3D, showing views around the site. These can either be presented as fixed 2D plans or using virtual reality (VR) the viewer can move virtually around the site.

Finalising the design

As a design gets closer to completion it can be increasingly difficult to identify what is wrong. At the early stages the problems are fundamental, but as these are corrected the remaining problems can be harder to define. The client may never be able to get to the point where they are 100% happy with the design, as there may be constraints that can't be mitigated or the budget may not be enough to fulfil their ideal scenario. Discussing this with the client before the design process starts can help make this later stage easier to manage, exploring which aspects of the project are their priority and if there are areas where they might be willing to compromise if needed. For a private project with no set timescale the client may be prepared to work until the design is as close as possible to their ideal, but for a commercial project with a constrained budget and tight timescale they may have to be more pragmatic.

However keen our client is to reach their ideal we can't continue revising for ever, so at some point we need to reach a resolution. We want to try to ensure that what is in our head is as close as possible to what is in our client's head, but no design will ever be perfect.

The technology used to represent our design can be as simple as a pencil sketch or as complex as a 3D-rendered walk through. There is skill in balancing the appropriate level of detail with careful use of the time allocated for the project. Over-delivering, such as creating complex plans that exceed the need for the workstage, means a reduction in profit. A fully coloured plan might be needed for a public consultation, but a simple line drawing may be all that is needed to make a decision on the location of a feature. Whatever method

GOOGLE EXPEDITIONS

The cost of the technology needed to create a virtual representation of a landscape continues to decrease, meaning that it is now within the reach of smaller organisation. An interesting example is the Google Expeditions project, a virtual reality app aimed at the classroom.[5] The app uses a tablet and mobile phones to allow teachers to take students on virtual field trips to sites as diverse as coral reefs and Machu Picchu. Using the tablet the teacher guides the class through a range of 360° and 3D images, highlighting points as they move through the virtual environment. The mobile phones are fitted into holders to create virtual reality viewers, as a low cost alternative to virtual reality headsets. Holder designs include foldable cardboard options.

In 2018 the Expeditions project was trialled in UK schools, to allow pupils to try the technology at no cost.[6] The project links to the Google VR programme, which connects to Google Street View and Google Earth. The system uses live-action imagery, but the same technology can be used to view computer-generated content, such as the output from CAD software.

The fact that virtual reality experiences can be created in the classroom using easily available technology means similar techniques are now becoming within the reach of even small landscape practices. Whether we use the technology to record site visits with 360° video or to walk a client around our designs, technology can help us to create better designs.

we use it is still only a means to an end – it is a tool that allows decisions to be made.

However involved we feel in a design and however much of our thinking it has occupied, it is always the client's project. This sounds a clear concept to grasp but with a long project or with a lack of direct contact with the client this fact can be overlooked. They set the parameters and we find solutions that work within those parameters. As the design progresses it can be easy to drift away from the original need, driven perhaps by functional need or ego. As the design stage closes we should to double check that we have really met the client's original need. We take their concept and add our own ideas and experience, but it is never our project.

THE PROJECT

Discussions during the creation of the project brief should have determined whether the project will be designed to an accredited standard. If this decision has been taken then there may be constraints on the materials or processes that can be used, with a knock-on impact on design decisions. If the project isn't being designed to an accredited standard the client and the project team may want to raise their standards higher than simply complying with legislation and look at the environmental and societal impacts of their project.

Materials and sourcing

The choice of materials is dictated by a range of factors, including cost, performance and availability. When selecting materials we also need to think of their long-term impact through production, maintenance and disposal. Landscape projects may potentially have a long life span but there still may be a point where the site is cleared and a new scheme created – what will happen to the materials we have selected, and have we passed an environmental problem forward rather than dealing with it now?

Some accreditation standards, such as the Living Building Challenge standard, have strict criteria for materials.[7] The limited range of landscape materials that currently meet such standards can limit design choices, but can also encourage creativity. Working with no environmental constraints allows far greater choice but may have unacceptable impacts. Agreeing within the project team what materials and processes are acceptable clarifies the standards the whole team will work to.

A client may want to review the project against their Corporate Social Responsibility (CSR) policy, ensuring that the project isn't at odds with the policy's aims. Clients who are cautious about reputational risk or who have had supply chain scandals may require greater scrutiny during product selection. This could cover the source of materials, working practices, country of origin, safety, environmental impact or even links to criminal activity. The complex supply chains typical of the construction sector can make it hard to trace materials back to their source elements. We can, however, start questioning suppliers and demanding more information about the products we specify and

Fig 4.2 Non-repairable plastic rail in an exposed location which has sagged in the heat and set in that position – the correct choice of materials to withstand site conditions is an important part of our work, 2013, Didcot, Oxfordshire

Fig 4.3 Inexpensive materials can work well if carefully detailed – concrete steps in London 2012 Olympic Park, 2012, Stratford, London

how they have been created, and push for greater transparency and improving standards.

We can't predict which materials will be banned or which working practices will become unacceptable but we can monitor opinion and research. One useful reference is the Candidate List of Substances of Very High Concern (SVHC).[8] However, even if a manufacturer can provide a list of components and their source, understanding the exact chemical make-up of the products can be a challenge for non-chemists. The International Living Future Institute (ILFI) maintains a Red List of the 'worst in class materials prevalent in the building industry', looking at the issue of bio-accumulation in the food chain and the impact on construction and factory workers as well as the impact on the environment.[9]

The Institute also supports the Declare labelling system, that states where a product comes from, what it is made of and where it will go at the end of its life.[10] At the time of writing there are very few Declare landscape products but it is a useful target to encourage manufacturers to work towards.

As a project team we need to work with the client to agree what materials are acceptable within our scheme. Researching the topic and making sure we understand the implications of our work means we can set our own limits on what materials and practices we are happy to support, and in what circumstances. We might be happy to specify a product with some adverse impacts if it is long lasting, requires limited maintenance and can be reused at the end of the project, but would object to the same material being used for a temporary show garden that only exists for a few weeks. Are there materials or practices that we would never agree to use? Do we make this clear to clients before we are appointed? Would we resign from a project if the client insisted on their use?

The circular economy and end use

Good design can substantially reduce waste, reducing environmental impact whilst saving the client money. Waste is often a hidden cost on projects – part sections of materials disposed of unused, existing materials with intrinsic value cleared from site or fees paid to send materials to landfill.[11]

Even at the very start of a project end-of-life issues should be considered. Designing with the end in mind should be standard practice even for such (hopefully) long-lived projects as landscape schemes , ensuring that we aren't passing a problem into the future by our choice of materials.

The traditional model of make > use > dispose is unsustainable. We need to keep resources in use for as long as possible, gain the greatest value from their use and recover or regenerate products at the end of each service life. The circular economy is a concept that aims to address this short-term approach to products and materials, assessing the whole-life cost rather than just the purchase cost. The concept aims to address the mismatch between the timescale the consumer works to with that of the supplier.

For a landscape project there may be different end-of-life timescales for different elements – short-lived shrubs, street furniture or lighting that may need replacing or repairing long before the end of use for the entire site. How the end of life of each element will

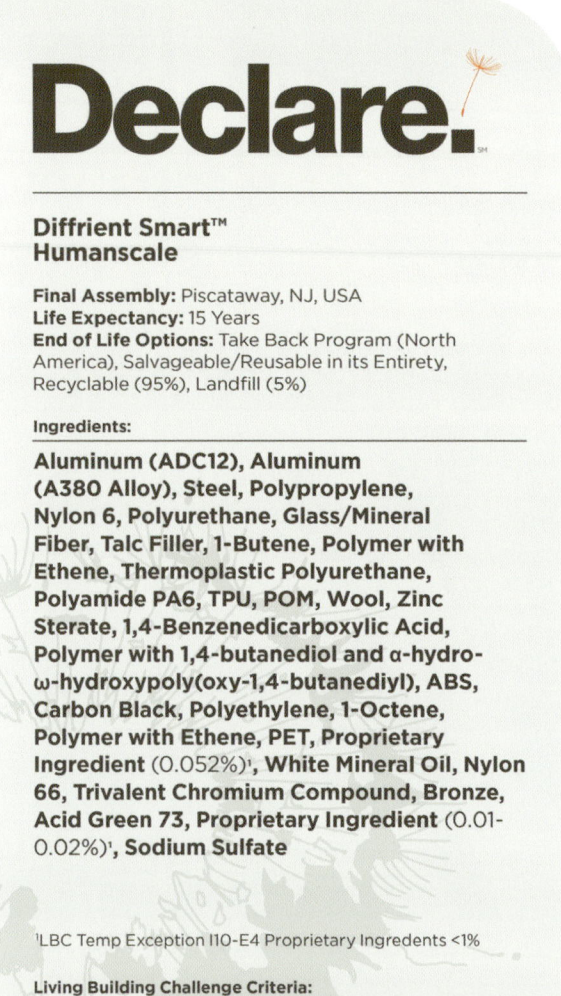

Fig 4.4 Declare label for Diffrient Smart™ Humanscale chair, one of the first Declare products in the UK – hopefully more products, including those in the landscape sector, will follow this innovative example

THE CIRCULAR ECONOMY

A good example of the mismatch between service timescales is domestic washing machines – consumers opt for cheaper, more affordable machines that have a limited warranty and are less energy efficient while, manufacturers can make machines that are more durable and more efficient but unaffordable for many.

Using a circular economy model the manufacturer would charge the customer on a usage basis, either paying per wash or leasing the machine for a longer period than a typical warranty. This spreads the cost over a longer period, covers all servicing costs and allows the manufacturer to recover the value in the machine. The manufacturer has an incentive to design repairable machines which can be upgraded and can be potentially leased numerous times.

Dutch architect Thomas Rau of RAU architects applied this principle when he wanted to fit out their Amsterdam office. Working with the lighting supplier Philips, Rau decided he wanted to buy 'light as a service' rather than an installation. Instead of a one-time purchase Rau employs a 'pay per lux' model, with Philips retaining ownership of the materials, providing maintenance and recovering components at the end of their service life. The bespoke lighting system is optimised to respond to natural daylight and occupancy levels.[12]

Both these case studies are from the Ellen MacArthur Foundation. Founded by Dame Ellen MacArthur, the fastest solo sailor to circumnavigate the globe and the UK's most successful offshore racer, the Foundation provides guidance for designers as well as case studies. Their website https://www.circulardesignguide.com/ provides online tools and free workshop templates to allow designers to learn more about the concept of circular design.

be dealt with should be a consideration during the design stage.

Including elements that can be repaired rather than replaced may be initially more expensive but may reduce the whole-life cost. Some materials, such as steel, natural stone paving or slate tiles, are so durable that they can be reclaimed and reused on future projects. The Living Building Challenge encourages this approach, promoting an end-of-life phase, rather than assuming items will be recycled. Depending on the type of material recycling is not always a good end-of-life solution, as each time the material is recycled it typically degrades in quality and is simply a delayed route to landfill.

A client with short-term aims and no long-term relationship with the site may not want to take this approach but as professionals we should at least take some responsibility for how our work will impact on future generations.

Off-site manufacture

Manufacturing buildings off site, usually in a factory, allows a higher quality to be achieved and removes the need for the building to withstand the weather as it is constructed. It also allows a different labour market to be utilised from that of the building location, moving the work to areas with low levels of development but high levels of skills. Off-site

QUEEN ELIZABETH OLYMPIC PARK – ASSET DISPOSAL CONTRACT

A central part of the London 2012 Olympic bid was the sustainable design which transformed the site post-games into a district scale regeneration site. This transformation stage required temporary roads to be removed, pathways and bridges narrowed and land use changed. This transformation stage could have resulted in significant waste but careful design meant this was reused. Structures were designed to make them simple to reduce in scale, with timber decking on bridges screwed rather than nailed or glued to allow for easier removal of bridge sections and temporary bridges were moved to permanent new locations.

An Asset Disposal Contract was set up to manage material reuse post-games, with assets sold or shared with the local community. Examples of reuse include:
– The reuse of the warm-up running track by British Athletics – the track was laid without tarmac allowing it to be lifted back up in lanes.
– Lamp columns donated to a local skate park.
– Timber tree planters – timber removed from temporary elements of the bridges has been used to create planters that can be moved around site as a short-term feature as the site is developed.

The Contract, whilst successful, did highlight some of the issues involved in disposing of assets, such as the care needed when dismantling items, the space to store them, creating agreements with community groups and getting assets to groups with limited resources. However, it does demonstrate what can be achieved if final use is considered from the outset.

manufacture is predominantly focused on buildings but there are elements that are relevant to landscape architecture.

Geo-tolerance

Whilst some components of our schemes can be manufactured off site and sections within a site such as pitches or play areas can follow a standard layout, one feature of landscape architecture is the level of geo-tolerance. A building may be able to be sited in a range of locations and still suit the conditions, for example standard housing types for housing developers, but a landscape scheme is only correct for that very specific location. Factors such as soil type, climate, altitude, aspect, exposure and the shape and topography of the site mean that every single scheme is bespoke. The mix of plants we have selected might not survive on a more windswept location close to our site, or on a different soil type. This specificity can be hard to get across to colleagues who are used to working with pre-made components that can tolerate a range of conditions.

There are landscape architects who design schemes without visiting the site, a fact usually given away by the misinterpretation of the site survey, but this unprofessional practice ignores the myriad factors we need to assess when completing our design, the factors that are unique to that site. There are some exceptions, such as sites that are too dangerous to visit, where we might need to rely on third-party surveys or remotely sensed information, but these are a rare exception. Some factors a landscape architect might need to consider include:

- soil – type, acidity/alkalinity, drainage, health, contamination
- climate – wind, heat, exposure
- aspect – slope, light levels, frost pocket
- noise – natural, manmade, positive, negative
- prevailing wind
- land use – historic use and potential contamination, future uses
- flow and circulation
- context – cultural, historic, setting within landscape
- light levels
- climax habitat type
- visual impact.

The sites we work on are unique, and require careful analysis to ensure that our design meets the constraints of that site. The materials and practices we choose for our projects have implications far wider than the scheme itself. Considering these at the design stage means we can minimise the long-term impact. Questioning our existing techniques and taking a long-term view should make us face up to the whole-life cost of the project. Our clients may not always wish to address the issue but we can at the very least highlight the potential consequences of their project and advise them of their options.

Fig 4.5. Off-site manufacture is not a new concept – columns were carved from the rock in quarries and then moved to site, Asclepeion of Kos, Greece

THE PROJECT TEAM

A project team that performs badly is unlikely to produce a successful project. What counts as poor performance can be hard to define, especially if different members of the project team have different expectations. For some team members it may be as basic as not returning emails or calls, for others it might be achieving an industry award or not ending up in court.

Not understanding the complexity of other team members' work can cause issues – not realising that what appears to be a minor change has knock-on effects and isn't the work of a moment to resolve. Symptoms of a poorly performing team might include missing agreed deadlines or not responding to queries, but can also include more subtle signs such as withholding information or not supporting other team members in discussions with the client.

Differing design styles and approaches to fulfilling the brief can cause problems, as can poor communication.

Collaboration and team working

For less complex projects we might not need to work as part of a team, dealing directly with the client, but even for small projects other professionals such as ecologists or surveyors are usually involved. Deciding who is part of the project team isn't always clear cut. The landscape architect can assemble and lead the team, ensuring our role is central to the project. In other scenarios the landscape architect can be seen as a sub-consultant who doesn't need to be involved in the core team. Being peripheral to the project team can mean that you aren't given the full background to decisions or are overlooked in decision making.

Taking the time to meet the project team, ensures the team know who you are and the role you are fulfilling. It also allows you to understand the roles other professionals will take, and who to contact if issues arise.

Team selection

Creating a team that has the right professional skills, understands the client's intentions and works cohesively is an important step in the success of a project. It is possible to create a good project with a disjointed team but it is likely to be an uncomfortable process, with internal conflicts potentially hampering progress.

For small or short-term projects these issues can probably be tolerated and time spent analysing the make-up of the team is misplaced, but for projects with timescales in years assessing the dynamics within the project team can be worth considering. Appointments are made based on professional skills and experience but how well each person works within a team can also be considered. An understanding of personality types and communication styles can help identify relationships that might be problematic, and perhaps even help in the selection of team members.[13] Creating a good project team isn't just about all team members getting on – research shows that some degree of conflict can help improve performance – but it does require a degree of trust, and at the very least civility, to be effective.

Setting up systems that encourage good communication and allow clear decision-making processes can help the team function effectively but there is always an element of luck involved when putting together a team. Someone who normally performs well in a team may become defensive if their job is at risk, or a normally enthusiastic person may be the opposite when personal circumstances, such as ill health, leave them tired. We can never assume a team will function well, and a seemingly well-performing team may fall to pieces if a serious problem arises. Conversely stressful situations can bring a team together, as long as all parties work together and divisions aren't created.

When selecting a team reputation is an important factor – not just the reputation of the individual, but of their team, their organisation and even of their profession. Care needs to be taken to test that reputation. An organisation may have been the subject of an unfounded media story, or a firm may spend substantial amounts of money marketing their skills to create a falsely inflated impression of their track record. If we start a project doubting a team member's ability we may be less trustful of their views, based solely on our perception of their reputation. A good reputation at an organisational level, where the values and standards are consistent and well-understood, means that that trust is transferred to all the individuals within that organisation.

Teams that need a high level of trust, such as military teams, rely on high levels of interaction to help them develop an understanding of each individual's capabilities.[14] Living and working together, often away from friends and family, allows them to develop a shorthand that they can use in high-pressure situations. Whilst construction project teams shouldn't need to work in such close proximity, or have such high stakes decisions to make, interactions can help team members understand each other's capabilities and help build trust.

Recreating a team that previously worked well can help but isn't a guarantee of success, as biases can set in where assumptions aren't questioned, or a sense of complacency can develop.

Team working may have potential pitfalls, but working as part of a creative and talented project team can be one of the most rewarding aspects of our work.

Team ethos

A team doesn't have to have the same level of trust as a military team but standards and values must be reasonably well aligned for the group to work effectively. Like a military team there must be a clear idea of purpose, provided by the brief, but there are less tangible factors such as a mismatch in values that can impede progress. Knowing our own personal, organisational and professional standards means we know what is and isn't acceptable.

Having these values set out, such as a code of professional conduct or a company policy, makes it clear to the rest of the team what values you will adhere to. Terms of appointment can also include basic values, such as the suggested 14-day payment terms in 'The Landscape Consultant's Appointment', published for members of the UK Landscape Institute, but unanticipated issues can still arise. There may be a difference of opinion as to an acceptable quality of work, or differing attitudes as to the priority given to the project.

Team ethos can include how fairly meetings are run, how work is allocated, the demands placed on team members, who has access to the client, the level of equality within the team or the level of honesty.

Bias in teams

Teams may feel that they are working in the best interests of the client, but they need to be wary of cognitive bias, a situation where errors in decision making occur as a result of sticking to unfounded preferences or beliefs. Biases that a project team need to consider include:

DEVELOPING TEAM ETHOS – SERVE THE STORY

When looking at how we can improve the performance of teams we can take lessons from seemingly unrelated sources. One example is a technique used in improvised comedy. When a group is improvising a routine, in the pressured environment of a comedy stage, there are rules that make the performance a success. No aspect of the routine has been planned, so each performer must trust the other performers implicitly. The technique asks that each person serves the story – if one performer sets up a situation that the next performer can do nothing with, then the whole team fails, so each action or line has to support the overall story. They may want to show off their quick wit or settle a score with a fellow comedian, but the routine only works if they all stick with this underlying principle. Writer, communications trainer and comedian Ryan Millar uses the concept when running workshops.

'I've had participants (when playing in a circle) get visibly and vocally irritated at being forced to use prepositions repeatedly. I've had others panic when it comes up to them to make a decision. They either fret that their contribution isn't enough, or conversely that too much is being asked of them. Both of these worries privilege the place of the individual. But, in these circumstances, nobody really cares about the place of the individual: they really truly don't. They just want a good story.

If that means you only say words like "the" and "an", then so be it. Your "poor me" attitude helps nobody and does nothing. The only metric for determining the success or failure is this: is the story any good?

As long as it is, that's all that truly matters. And if you're given the responsibility of coming up with something that influences the direction of the story... well, again: it's not about you. It's about the story. So say whatever makes sense for the story. Job done.

This shift in perspective is small, but significant. For some it's a slight and subtle shift, or even a confirmation of something they already understand on a profound level. For others it is a challenge to their ego and identity. Regardless of the difficulty level, the mantra is the same: Serve. The. Story.'[15]

Swap the word story for project and you have a useful convention for project teams.

Confirmation bias – The tendency to favour information that supports a pre-existing belief, in turn ignoring information that conflicts with our view. This means decisions are not objective or based on the full range of information available.

Optimism bias – The tendency to be over-optimistic about project parameters including costs, duration and benefits delivered. Over-optimistic estimates lead to targets that are undeliverable. The HM Treasury document 'The Green Book – Guidance on Appraisal and Evaluation' has useful guidance on this bias, including a table to work out an optimism bias adjustment for cost and work duration in civil engineering and building projects.

Self-serving bias – The tendency to present information to maintain or enhance self-esteem, such as overstating the level of progress or capabilities. This means decisions are made based on inaccurate information and problems can be understated.

Information bias – The tendency for people to continue seeking additional information before making a decision, even if no further benefits are gained. This can delay decisions and incur additional costs.

In-group bias – The tendency for people to favour members of a group they belong to. At the simplest level this bias can prevent groups collaborating outside their immediate circle, but at the most extreme level it can manifest as discrimination or prejudice.

Recency bias – The tendency to place more emphasis on recent events, overlooking longer-term trends.

Groupthink – The tendency to be influenced by the opinions and actions when operating in a group. This can lead to a false consensus as team members form a cohesive group where ideas aren't challenged.

Bias blind spot – The inability to identify and compensate for biases.

Planning fallacy – The tendency to make overly optimistic predictions about time, cost or likelihood of success when planning a future task or project. This can be motivated by the desire to win bids or wanting to see a project complete.

Sunk cost fallacy – The tendency to make decisions based on past costs (time, effort or money) that have already been incurred and cannot be recovered. This leads to an escalation of commitment, causing us to continue when aborting or changing direction might be a better option.

Overcoming bias

Cognitive bias can be difficult to identify, as the elite-sports psychiatrist Professor Steve Peters identifies in his book *The Chimp Paradox*. Peters, who works as a performance coach with athletes including the British Cycling Team, argues that we have a human brain that can process problems rationally but this often at odds with our chimp brain, which jumps to conclusions based on patterns that we believe are correct, sometimes based on inaccurate assumptions.[16] These behaviours are linked to our evolutionary success, and ensured our survival as hunter-gatherers, but are less helpful to us in the modern world especially when we are required to make far-reaching, rational decisions. Heightened emotions or stressful situations, when our chimp brain comes to the fore, can increase the risk of bias.

Pre-mortem – An exercise where project teams imagine their project has failed, and work backwards to look at all the reasons this could have happened. Where possible, realistic contingencies for these scenarios are included in project plans. This technique also helps create an environment where the likelihood of set backs is recognised. A pre-mortem might identify that the time of year proposed for tree planting was at a time of year where plant stocks could be low. To mitigate this the team might reserve plant stock earlier than planned, removing the risk of not being able to source the right material at the right time.

Red teaming – Using an outside team to challenge the team's assumptions and plans. The red team have no link to the project so are unmotivated by the effort and investment taken to date to progress a project. This technique is used by sectors such as the military and aviation. A red team may be able to spot risks that the project team had overlooked, or help them realise that a minor risk is taking too much of their attention.

Decision trees – Decision trees set up a pre-defined pathway that users work through to make decisions. Each step has a yes/no option, with criteria for each. The aim is to remove emotion from the process. Using a decision tree allows the user to demonstrate their decision-making process. Decision trees are commonly used in medicine, as they provide a summary of current practice and can help offset the impact of stress and tiredness. A good example is one which helps doctors decide if a child with a head injury needs a head scan. Decision trees can be used to help decide which projects are worth taking on, how a team is performing or when it is worth abandoning a project.

Roles

Unless we are the only designer on a project we will need to collaborate with others to produce a design. To collaborate effectively we need to be clear about our roles. Something as simple as deciding a kerb line may involve a number of professionals, all looking at different aspects of the decision. Working out who needs to be consulted, how these decisions are made and how consensus is achieved is a valuable part of the collaborative design process.

Drawing up a responsibility matrix can help define who is responsible for which aspect, and who needs to be consulted on that issue. As landscape architects it can be frustrating to discover that an underground service has been rerouted to pass through a tree pit, because the person making the change didn't realise the potential impact on the landscape design. Defining roles and making it clear what we need to be informed about can help reduce the risk of such errors.

As the project progresses it is worth reviewing if roles are being fulfilled, or if they are still correct if the project has changed substantially in the intervening time. Are roles being adhered to? Is someone intruding on our work, or we on others? Are we really collaborating or are we competing?

Access to the client

If one team member is given responsibility for reporting to the client, perhaps as lead consultant, the rest of the project team need to be confident that the situation is being fairly represented. Access to the client gives that person an advantage. If they are prone to self-serving bias, want to promote their viewpoint or cover up an error they may abuse that advantage. However, having a single point of contact with the client means that conflicting information is avoided and the most efficient use made of their time. The client needs a clear system of reporting, but this must reflect the situation as honestly as possible.

The case for collaboration

In the UK construction sector there is a consensus as to the benefits of working collaboratively, in part prompted by the findings of the Latham, Egan and Farmer reviews. Previously adversarial, with a language of claims and damages which highlighted that conflict was seen as the norm, the sector has moved towards a less combative approach.

Table 4.1 (Page 107) Design Responsibility Matrix to show project roles during design stages, based on the matrix in the RIBA Plan of Work Toolbox

DESIGN RESPONSIBILITY MATRIX

ASPECT OF DESIGN		2 - CONCEPT DESIGN DESIGN TEAM			3 - DEVELOPED DESIGN DESIGN TEAM			4 - TECHNICAL DESIGN DESIGN TEAM		
Classification	Title	Design responsibility	Level of detail (LOD)	Level of information (LOI)	Design responsibility	Level of detail (LOD)	Level of information (LOI)	Design responsibility	Level of detail (LOD)	Level of information (LOI)
Ss_15 - EARTHWORKS										
Ss_15_10_30	Excavating and filling systems	Landscape Architect	2	2	Landscape Architect	3	3	Civil Engineer	4	4
Ss_15_10_31	Earthworks filling systems around trees	Arboriculturist	2	2	Landscape Architect	3	3	Landscape Architect	4	4
Ss_30 - ROOF, FLOOR AND PAVING										
Ss_30_14_05	Asphalt road and paving systems	Landscape Architect	2	2	Landscape Architect	3	3	Landscape Architect	4	4
Ss_30_14_80	Unbound aggregate paving systems	Landscape Architect	2	2	Landscape Architect	3	3	Landscape Architect	4	4
Ss_37 - TUNNEL, SHAFT, VESSEL AND TOWER SYSTEMS										
Ss_37_16_65	Pond and wetland systems	Civil Engineer	2	2	Landscape Architect	3	3	Landscape Architect	4	4
Ss_45 - FLORA AND FAUNA SYSTEMS										
Ss_45_10_95	Vegetation control systems	Landscape Architect	2	2	Landscape Architect	3	3	Landscape Architect	4	4
Ss_45_30_05	Aquatic and wetland planting systems	Landscape Architect	2	2	Landscape Architect	3	3	Landscape Architect	4	4
Ss_45_35_05	Amenity and ornamental planting systems	Landscape Architect	2	2	Landscape Architect	3	3	Landscape Architect	4	4
Ss_45_35_30	Forestry, biomass, hedging and roadside planting systems	Landscape Architect	2	2	Landscape Architect	3	3	Landscape Architect	4	4
Ss_45_35_30_62	Pit-planted tree and shrub systems	Landscape Architect	2	3	Landscape Architect	3	3	Arboriculturist	4	4
Ss_45_35_45	Lawn and meadow planting systems	Landscape Architect	2	4	Landscape Architect	3	4	Ecologist	4	4
Ss_50 - DISPOSAL SYSTEMS										
Ss_50_70_85	Sustainable drainage systems (SuDS)	Civil Engineer	2	2	Landscape Architect	3	2	Landscape Architect	4	4

CLASSIFICATION
this uses Uniclass 2015, a universal classification for the construction sector, including landscape architecture, that uses a hierarchy of codes to classify information.

LEVEL OF DETAIL (LOD)
the degree of graphical information. Level 2 for a tree be a simple graphic with no species or size to, whereas Level 4 would be a graphic, perhaps including a cross section, with enough technical detail for tender and construction.

LEVEL OF INFORMATION (LOI)
the degree of non-graphical information. Level 2 for a tree would state a tree and perhaps the surface it is planted in, whereas Level 4 would include enough technical detail for tender and construction.

Collaboration and BIM

Collaboration is central to the basic tenets of BIM. The communication techniques, document management and methods of change control all encourage and require collaborative working. To ensure that all communications and decisions are open and shared they are saved in a single collaborative data space, the Common Data Environment (CDE). BIM principles, such as interoperable file formats, mean that each designer's work can be included in one 3D digital model, for the first time allowing the whole team to see their work combined in real time. This shared model can be used to detect clashes in the design, such as services running through tree pits or mismatches in levels. Working in BIM can be a challenge for landscape architects, as the components we use don't fit neatly into categories, and in the case of plant stock are not a predictable size and grow during the life of the project, but the potential gains are substantial and can help us be recognised as an equal member of the project team.

Another challenge is the reluctance to geo-locate digital models for buildings – there will be a point that the team work to, but this doesn't always link to a precise location such as an Ordnance Survey grid point. Anther common issue, not just with BIM, is orientating the drawing to match the building orientation, rather than keeping the original survey orientation, which is usually with north as up. This isn't a problem for most of the design team, but for the landscape architect it is a major issue as they need to pay careful attention to areas of sun and shade as well as placing the scheme in a wider setting.

However, BIM is still a useful process for landscape architects, with benefits as wide ranging as consistent layer naming in CAD drawings and dynamic drawing take off, with plant numbers and paving areas automatically calculated as the drawing evolves. The ability to visualise our design in three dimensions, allowing clients a better insight into our designs, is a major step forward for our profession. The book *BIM for Landscape* explains in more depth how landscape architects can implement BIM into their work.[17]

Design and project team meetings

The traditional approach for team communications and monitoring progress is to hold regular face-to-face meetings, either with the whole project team or just the members of the design team. The default is monthly meetings but ideally they should be scheduled so decisions can be made and work coordinated to meet the project timetable.

Face-to-face meetings are an opportunity to interact with other team members, but this needs to be tempered with the time taken, including travel, for what might be minimal contribution. Improvements in video-conferencing systems, including screen sharing, mean that attending a meeting remotely is now a practical and affordable option and is common practice for landscape architects working internationally. Not attending the meeting in person may mean you miss out on informal discussions outside of the meeting but if the meeting is well organised decisions are only made in the meeting and accurately reported.

MANAGING FEE INCREASES

Would you expect to pay extra for ketchup in a restaurant? Would you assume it was included in the cost, along with salt and pepper, or accept that it was an extra and pay? Most food items and their price are clearly laid out in the menu, with the cost for additional extras such as side dishes set out but there is a grey area for some items. Is tap water included? How long can you stay at your table before the next customer arrives?

We have similar dilemmas with our clients – we might think of extra items as an additional cost but they might assume it was included in the fee quote. Unless we have explained the scope of our quote there may be a mismatch in expectations. When we manage fee quotes we need to be clear what is included, what extra items are excluded and what the cost of those items would be.

Reviewing past projects and ensuring that adequate time was allowed in fee quotes for meetings, along with administrative tasks such as emails and paperwork, can help gain a better understanding of where the time was spent and help improve the accuracy of future quotes.

We also need to make it clear that changes in deadline may incur additional costs. We need to be careful that by explaining the fee increases for changes in timescale, costs we might see as a deterrent, we don't inadvertently encourage the client to choose this option. Incentives not to change timescales include non-financial factors such as not wanting to place additional pressure on a design team or understanding that a reduced timescale is likely to result in reduced quality. Once an option to financially compensate for the changes is introduced that option is then open to the client, and what we included as a deterrent becomes a price for urgent work.

Increased costs need to be explained to the client and agreed before the cost is incurred. When managing projects we might allow a certain percentage of spend over that costed in our quote, perhaps for a long-term client who we enjoy working with, but that needs to be a deliberate business decision rather than a gradual and accidental slide into loss.

Attendance at meetings is an easy and sometimes hidden way to make a loss on a project. If the project overruns the number of documents produced by the landscape architect might not increase but attendance at extra meetings not included in the fee quote can rapidly erode profit. Stating in the fee quote the number of meetings allowed for and a rate for additional items such as attendance at additional meetings can help, as can careful monitoring of time spent on a project, but it can be difficult to agree additional fees when urgent issues need resolving.

Procurement

As the Farmer Review showed, procurement in the construction sector has fundamental flaws. At its most basic the process usually involves issuing a tender, assessing the tender submission and then awarding the work to the lowest bidder. The client then has to fund any gap between the tender value given in order to win the work and the true cost of the project.

Contract terms help address this issue and set out procedures to resolve pricing disputes but ultimately much of the risk sits with the client, who once they have started a project are usually committed to completion. Innovations such as BIM, with the ability to automatically create detailed schedules that dynamically change if areas are changed, mean that more accurate schedules can be produced, in turn allowing more accurate costings. If a 3D model is created as part of the BIM process the design can be tested at a virtual stage, allowing the opportunity to check for clashes and for the client to check the scheme in detail before it is committed to site. One of the future anticipated stages of BIM includes real-time costing, known as 5D BIM. By including smart elements in drawings and models, items that have data attached to them so they can be identified, the cost of our designs could be estimated in real time.

Performance

Agreeing what constitutes good performance and the standards within a project team means that it is clear when these are not met. The client has the ultimate sanction – if they are not happy with a team member's performance that person or company can be removed from the project. This is a drastic step, and may have knock on implications such as delays or increased costs, but if expectations have been made explicit and all chances to explain or rectify exhausted then it can be the correct course of action. Persevering with a team that fails to perform, and ignoring the reasons behind this, is unlikely to result in a successful project.

A project team needs to be organised and coordinated to allow them to produce a single unified design. Understanding the client's needs, working well together and communicating effectively is a good basis for a successful project.

Collaboration between disciplines doesn't mean conceding our role or even always agreeing with the rest of the project team – it means building consensus and, usually, compromise. Our work can be as much about persuading the project team of the value of our work as it is about persuading the client.

WIDER SOCIETY

Our projects don't exist in isolation. Much of our work as landscape architects is either used by, or at the very least has an impact on, wider society. One of the pleasures of landscape architecture is that much of our work has a positive impact on wider society – public spaces, new habitats, liveable housing areas – with clear benefits. The public is not the client and may not be involved in the design process but as professional designers we need to consider the impact of our design on wider society. For some clients, such as public-sector organisations or charities, the needs of the local community might be central to the design. Other clients, such as developers of new commercial areas, may want to more carefully manage who uses the site.

During the design stage a clearer idea of the project emerges, allowing us to consult with outside organisations and those potentially impacted by our work. If our project requires planning permission or other consents, the impacts on wider society will be tested by outside parties. For community projects

a consultation process can be part of our remit, allowing us to test the design against the community's requirements and expectations.

Hostile design

Hostile design is a technique that discourages people from using a public space in certain ways. This could be benches that are deliberately uncomfortable, boulders under bridge spaces to prevent rough sleeping, spikes along sills or metal 'pig noses' along structures to deter skateboarders.[18] Some features are explicit, such as the spikes, but others are more implicit such as playing music that deters a certain age group or those wanting to sleep.

The concept of hostile design is not new. In Victorian London urine deflectors were installed in many of the alleyways popular with perpetrators. These low, angled shelves deflect the urine away from the wall and back over the feet.

Many of the scenarios where hostile design is introduced compensate for human vigilance, either by paid staff such as park keepers, caretakers or by the community itself. The original reason for the technique may have a practical basis, such as a reduction in crime or anti-social behaviour. However the measures are indiscriminate, making public spaces less welcoming to users such as people with disabilities, older users and children.

Fig 4.6 Metal pig noses along edge of low stone seat to deter skate-boarders, 2019, Frideswide Square, Oxford

Fig 4.7 Anti-urination structure in corner of doorway, 2018, Whitehall, London

Security

Part of our work as landscape architects is to ensure that the people using the spaces we design are safe. The level, visibility and appropriateness of security measures need to be carefully considered and all options explored. Security measures need to be proportionate – unnecessary security measures can inflate costs and a proliferation of overt security measures can act as an unintentional marker for areas of deprivation. In an affluent area the secure boundary might be a beautiful brick wall, or Victorian wrought-iron railings. In a less affluent area that same boundary might be created with weld-mesh panels or spike-topped palisade fencing, giving the unintentional signal that this is not a place to linger, in turn reducing natural surveillance.

'...that the sight of people attracts still other people, is something that city planners and city architectural designers seem to find incomprehensible. They operate on the premise that city people seek the sight of emptiness, obvious order and quiet. Nothing could be less true. The presences of great numbers of people gathered together in cities should not only be frankly accepted as a physical fact – they should also be enjoyed as an asset and their presence celebrated.'[19]

Jane Jacobs, *The Death and Life of Great American Cities*

In the 'Secured by Design' strategy for the London 2012 Olympic Games overseen by the Olympic Crime Prevention Team the aim was to 'deliver a beautiful

Fig 4.8 People sharing public open space in the late summer sun, 2018, Russell Square, London

venue and park underpinned by subtle but integral security planning measures'. Elements including CCTV, lighting, fencing and planting were designed to deter crime and ensure both formal and natural surveillance were possible. Along main routes planting was kept low, and fencing included clear strips to afford visibility.

According to the report 'Fortress Britain' highly visible security measures can raise our concerns over safety.[20] In certain circumstances that could be a deliberate intention, such as a military site or a border crossing where vigilance is required. However inappropriate levels of security can make urban areas feel militarised or suggest that there is a greater risk of crime. As landscape architects we need to strike a careful balance between deterring anti-social or criminal behaviour and creating welcoming and inclusive public spaces.

Consultation or exhibition?

When we share our work with wider society we need to be clear if we are running a consultation or an exhibition. In a genuine consultation the results of the discussion can potentially change the course of the design. It isn't too late to make amendments and the designers are open to the ideas made by those involved. If the design being shared has been fixed with no possibility of changes then it is an exhibition not a consultation.

There is a place for exhibitions, for example where a scheme has been agreed and the benefits and potential impacts need to be explained, where there is only one solution to a technical issue such as flood risk, or at the end of a consultation process to share the final design. However, tokenistic consultation where no change can be effected, or consulting on a trivial issue such as the colour or pattern of one element, reduces the credibility of our profession and patronises the communities for whom the impact of our work may last long after our involvement is complete. It also makes the work of the next organisation undertaking a consultation more difficult as the level of trust has been reduced.

A genuine consultation allows the design team to fully understand the context of a site, using the knowledge of those who understand it in greater depth, perhaps even better than the client. A new staff member at a local authority may not know that a site is used for a carnival, with complex access needed for heavy vehicles. A beneficial group might informally use a site. Talking to the community, either at open events or as individuals, can help gain real insight and allow us to create more sustainable schemes that aren't at odds with the needs of the wider area.

One of the best ways to hold a consultation for changes to existing landscapes is to consult on site. Standing and talking to people who use the space as they use it is a great way to gain a more representative impression than the traditional formal consultation event. It isn't always practical, such as with greenfield or remote sites, but setting up a table, offering a hot drink to anyone who will stop and chat and having unstructured one-to-one conversations without the comfort blanket of a formal display can help gain insight that a self-selecting group attending an off-site event could never reveal.

If it isn't possible to hold a consultation on site, then it should at least include locations that draw a cross section of people, such as shopping centres or health centres, or be as part of a larger popular event such as a local show or festival. Holding events in buildings that are not accessible or on a public transport network, at times when most people are working or are inconvenient for those with caring responsibilities, means that we only hear certain views that may not be truly representative.

Supply chains and modern slavery

The impacts of our work spread far wider than the site we are working on. The materials we specify can include elements from all over the planet, as do most of the manufactured products in our lives. Our smartphones contain rare earth minerals, many of which are sourced in conflict zones or parts of the world with poor working conditions.[21]

The complex supply chains typical of the construction sector mean that it can be difficult to track the supply chain of a product – unless the product we choose is accredited to a labelling system such as Declare (see Fig 4.4) we are unlikely to know the source of the individual components, the working conditions of those involved in manufacture or the environmental impact of production.

Poor working practices, the extreme extent of which is described as modern slavery, is a worldwide problem. In the UK the Home Office, the government department responsible for immigration, security and law and order, estimated that in 2013 there were between 10,000 and 13,000 victims of slavery in the UK.[22] The International Labour Organization estimates that 'three out of every thousand people worldwide are trapped in jobs into which they were coerced or deceived and which they cannot leave'.[23]

Modern slavery is work that is performed involuntarily and under the threat of any penalty. A victim of modern slavery has one or more of the following characteristics:

- forced to work through coercion or by mental or physical threats to them or their family
- owned or controlled by an 'employer' through mental or physical abuse, or the threat of abuse
- dehumanised and treated as a commodity or bought and sold as property
- physically constrained or have restrictions placed on their freedom of movement by the retention of identity papers
- receives little or no wages for work, or working to 'repay' a debt to the enslaver.[24]

Fig 4.9 Public consultation event with coffee cart to provide free drinks for participants; Thirlwall Associates for Sustrans 2014, Reading

Public-private space

Over the past decade in the UK there has been a move towards privately owned public spaces, often as part of the redevelopment of a site. Unlike public spaces there is rarely right of access, and the owner can set conditions that prohibit certain behaviours, such as holding public protests or taking photographs.

Partnerships with private developers can enable the provision of open space where public funding isn't available but it is worth remembering that the motive of a commercial landowner is to protect the commercial interests of their site, as opposed to providing a public service.

MODERN SLAVERY WORK BY THE CHARTERED INSTITUTE OF BUILDING (CIOB)

For many people in the UK the first time they became aware of modern slavery was in 2004 when the deaths of 23 Chinese cockle pickers working in Morecambe Bay was widely reported in the British media.[25] The workers had paid huge amounts for their illegal passage to England. On arrival their passports were taken and they were forced to live in cramped conditions.

After the tragedy legislation was introduced to try to address the issue, including the Modern Slavery Act 2015. The Act sets out the requirements for UK businesses to assess their supply chains, but only companies with a turnover of more than £36 million a year are required to report.

There may be no legal requirement for smaller companies to assess their supply chains but legal obligation should not be the only motivation – we risk our and our client's reputations by ignoring such issues as well as being unwittingly complicit in human rights abuses. In the construction sector the combination of complex cross-border supply chains, multi-tiered procurement and fragmented recruitment processes mean that exploitation by bonded or forced labour can remain hidden.

'With its labyrinthine supply chains, stretching over countries and continents, construction provides the perfect breeding ground for exploitation and human rights abuses.'[26]

In 2016 the CIOB produced the report 'Building a Fairer System: Tackling Modern Slavery in Construction Supply Chains' as part of a wider campaign to eradicate unfair labour practices.[27] Written with the support of a wide range of organisations, including Amnesty International, Engineers Against Poverty and Office of the Independent Anti-Slavery Commissioner (UK) the document includes an explanation of how modern slavery occurs in the construction sector and sector specific case studies. One case study covers the supply chain of a UK-based hard landscape manufacturer who discovered child labour within their supply chain and their work to address the problem.[28] The report also includes a directory of helpful organisations, websites and resources.[29]

Who are we designing for?

Whilst creating the brief we should have discussed the expected users of the site. For a private scheme we might know the individuals personally, allowing us to create a landscape that suits them as closely as possible. In most cases we won't yet know the potential users of the site, so we tend to design to the median of the population. It won't be consistent throughout the whole design but for each item we include, such as a bench or a handrail or even a litter bin, there will always be a height or build for which that object is optimal.

As people move through our schemes, there will be users for whom the whole design is optimal – the surface doesn't cause them balance problems, they can cross the road in the time allowed by the traffic signals, the lighting levels are ideal and there are enough benches at the correct height.

STOCKHOLM – A CITY FOR EVERYONE

In 1999 the Swedish capital city of Stockholm set the target of becoming the world's most accessible capital by 2010. This target has been further extended, with the aim of becoming 'a hub of an accessible and safe region with no social or physical barriers' by 2030.[30] Improvements have been made to public transport, public spaces, retail and accommodation as well as information campaigns. The underlying guidelines for the programme are the UN Convention on the Rights of Persons with Disabilities.[31]

Convention on the Rights of Persons with Disabilities (2006)

Article 1 – Purpose

The purpose of the present Convention is to promote, protect and ensure the full and equal enjoyment of all human rights and fundamental freedoms by all persons with disabilities, and to promote respect for their inherent dignity.

Persons with disabilities include those who have long-term physical, mental, intellectual or sensory impairments which in interaction with various barriers may hinder their full and effective participation in society on an equal basis with others.

Since the targets were set work has been undertaken across the city, with improvements to public transport, street layout and public open space.

Specific improvements include:

– **Pedestrian crossings** – crossings have a kerb and a ramp, giving visually impaired users a clear edge to navigate to and flush access for wheelchair users. The crossing control makes a quiet knocking noise so it can be found by the visually impaired, beeps to acknowledge it has been set and then emits a rapid beep as the traffic is stopped. Studs are placed in the road surface to guide visually impaired users.[32] By 2010 5,200 crossings had been converted to the Stockholm model.[33]

– **Pavement design** – The use of tactile paving is consistent, with marker paving across the whole pavement where there is a crossing, and along the line of the pavement in the approach to crossing points.

– **Stepped access** – High contrast markers were added to the first and last step to help those with a visual impairment, outdoor lifts were installed where needed and steps removed where feasible.

– **Public toilets** – People with some medical conditions need to have easy access to a public toilet. The provision of toilets was assessed so that those with such conditions could be confident when visiting the city.

Fig 4.10 Stockholm crossing – dropped kerb and stepped kerb with tactile paving, 2005, Stockholm

Fig 4.11 Choice of materials can affect who can use a space – Corten steel as a surface becomes a slip hazard in wet weather, 2015, UK

MASLIN-CLINTON ENVELOPE OF NEED

This potential for exclusion has been identified by Steve Maslin and Michael Clinton of the Schumacher Institute for Sustainable Systems, who have developed a theory for accessible design based around principles inspired by Clinton's background in aviation.

Maslin's experience as an architect and access consultant, specialising in enabling mind friendly environments, and Clinton's work as an aeronautical engineer and researcher, allowed them to develop their concept.

In aircraft production the designer seeks to determine the performance envelope of the aircraft that they are designing - the limits of the design such as airspeed, load and the conditions that will cause the engine to stall or otherwise come to grief. The Maslin-Clinton's Envelope of Need uses these same principles to help with the design of internal and external spaces. The concept looks are the extent of the "envelope" can be made, and how the needs of as broad a range of users as possible can be accommodated in a setting.

In landscape design we don't have clear-cut tests of performance, but we can use the concept of the Maslin-Clinton Envelope of Need to test our designs. Which users are within the envelope? Who is excluded, and has every option to correct this been explored? Challenging topography, low budgets and short timescales can all be used as excuses to create schemes that exclude some users but good, thoughtful design can overcome many of these limitations. One case

Someone who finds walking easy, is of average height, has perfect sight and hearing and doesn't struggle with their balance will find most spaces easy to negotiate.

Being outside the optimal range for a scheme, even perhaps being left-handed in a predominantly right-handed world, can make moving through a space less efficient. Handle layouts and ticket machines are often designed for right-handed users. Each step away from the optimum design, such as the need for frequent rests, or a visual impairment, makes the experience of the space less satisfactory.

Our aim for any public landscape should be to remove as many barriers as possible.

For each sense a list of considerations are given, against which any design should be assessed.

SENSE	NEEDS THAT SHOULD BE CONSIDERED
Mobility	Wheelchair users, walking aid users, those with balance issues, people who can only walk a limited distance without rest
Vision	No sight, tunnel vision, issues with low-contrast lighting
Hearing	No hearing, tinnitus, balance and hearing issues
Metabolic	Temperature, need for food, access to toilet facilities
Neurological	An environment that does not trigger unwanted neurological symptoms, such as visual interference from patterns in paving

To address each need there is a design response, some of which apply to a number of needs.

DESIGN RESPONSE	DESCRIPTION	SOLUTIONS
Personal logistics	Information required for users to plan their visit to a site including travel, welfare facilities, access arrangements and details of any pre-booked support	Pre-arrival information provided in digital or paper form, such as descriptions, photographs or 360° videos of routes. The information can include descriptions of surfaces, gradients, distances and the proximity of facilities to allow the user to decide if the site is accessible to them. Providing a choice of routes gives users options
Legibility	Signage	Clear signage that works for people with visual impairments and neurological conditions
Clarity	Including clarity of routes, vistas and wayfinding, as well as the visual and auditory clarity of the space a user is in	It is part of design to vary these to allow different experiences but spaces used by all visitors, such as catering spaces and information points, should have a level of clarity, such as lighting and acoustic conditions, that suits all users
Psychology of the environment	Neurological considerations, meaning and metaphor	Ensuring that there are no features that deter users from using a space, such as features that trigger unwanted neurological symptoms, or negative visual cues such as offensive insignia or signage
Ergonomics	Designing spaces that fit the people who use them	Understand the needs of potential users and account for these in the selection of items such as street furniture, toilet provision and handrails

Table 4.2 Maslin-Clinton Envelope of Need

LONDON 2012 – ACCESSIBILITY INFORMATION FOR VISITORS

For visitors with a disability planning a visit to a new site can be complex. Having to book assistance rules out spontaneous trips and requesting help can be awkward. Considering the issues that a visitor with a disability may need to address and providing as much information as possible should be part of the accessibility strategy for any site. Trying to classify or grade a site isn't always helpful – each visitor has a unique set of requirements so it is better to provide detailed information that allows the visitor to make their own judgement about how accessible the site is to them.

One of the targets for the London 2012 Olympic Games in London was 'to remove attitudinal and environmental barriers that create undue effort, separation or special treatment. This will enable everyone – regardless of disability, age, gender or faith – to participate equally, confidently and independently with choice and dignity.'[35]

The information provided on the London2012.com website gave visitors clear and detailed information to allow them to work out the logistics of their visit and to arrange any assistance they might need. The Accessibility section of the website gave information on a wide range of topics including:

- parking – visitors could book a Blue Badge space via the London 2012 website[36]
- accessible travel – additional services were provided for people with a disability, including accessible shuttle buses and accessibility improvements to some public transport services
- games mobility – assistance available to spectators within venues, including wheelchair loans and guiding for visually impaired people
- toilet provision, including Changing Places (adult changing) toilets
- assistance dog spending areas
- how spectator information would be provided, including accessible formats and audio description and commentary.

sometimes given for not needing to address these exclusions is the lack of demand. Why improve access when no one with access needs ever visits the site? This misses the point that there is often a hidden demand, an unmet need, that will only be revealed when the barriers are removed.

If we fit neatly into the performance envelope for most sites, we might never realise the difficulties that other users may encounter. Our own experience, such as caring for children or travelling with a wheelchair user, may help us understand some of the issues, but unless we talk to potential users or our circumstances change we might never realise the barriers we have inadvertently included. In many countries legislation sets out the minimum standard for access, such as the the Equality Act in the UK, or the Architectural Barriers Act in the US, but these only set out the minimum required. The Convention on the Rights of Persons with Disabilities is a helpful starting point in the absence of state legislation.[37]

Considerations include:

Signage – Does the signage work for those with a visual impairment, or dementia? Are pictorial signs more appropriate?

GOOGLE MAPS –
A CAMPAIGN FOR
ACCESSIBLE ROUTES

In 2017 UK student Belinda Bradley started an online petition to ask Google to create wheelchair-friendly maps, after struggling to travel around London with her mother, who is a wheelchair user. The petition gained massive support, and after the campaign was picked up by national newspapers Google responded.[38]

Belinda wanted users with access needs to be able to use the same software as other travellers, rather than have a separate app, and has worked with Google to create a solution. Still in its very early stages and at the time of writing only available in London, New York, Tokyo, Mexico City, Boston and Sydney, the idea is that Google will work with local transport providers to improve information, including Street View imagery inside stations. The work will also be supported by members of their Local Guides community, volunteers who record data about their community in return for small rewards.

Anyone with a Google account can become a Local Guide, so access information can be added for any landscape scheme if listed on Google Maps.

Wheelchair accessible routes are not shown by default, and need to be selected via the options settings, so there is still work to be done to publicise the option, but it is a great example of user pressure resulting in change.

Fig 4.12 Universal design is of limited use if it isn't maintained – broken and weedy tactile paving, 2013, Didcot, Oxfordshire

Surfacing and pattern – Does the design we've chosen cause visual disturbance to those with a neurological condition?

Rest points – Are there places to rest at regular intervals? These can be seats but it can also be useful to include features that are not overtly resting points, such as the edges of planters or perching posts. These provide people with balance issues with a discreet place to steady themselves, or a place to briefly pause if walking is difficult.[39]

Route descriptions – Is it clear what the conditions are as a visitor begins a circular route, for example in a park? Providing information such as distance, surface, gradient, light levels and rest options means a user can decide if a route is suitable for them. This can be provided via signage, printed leaflets, downloadable leaflets or downloadable audio files.

Cultural signals – Are there subtle signals that suggest the landscape is only for certain types of user or that could be seen as offering offence?

HEALTH AND WELLBEING

There is a general consensus that time spent outdoors is good for physical and mental health, and as designers of landscapes we often work to promote the value of quality open spaces. However, in some cases the rationale can be simplistic or unproven, which risks weakening our argument for access to open space. The famous and much-cited 1984 study by Professor Roger Ulrich in a suburban Pennsylvania hospital suggested that having a hospital bed in sight of a natural setting 'might have restorative influences'.[40] This important paper does show a difference in recovery times for patients with different views but the patients only viewed the setting from a hospital bed, so doesn't provide an argument for the provision of accessible open space. Interestingly the comparison is only for the times of year the trees had leaves, and only for patients without serious complications.

The paper finds that 'the conclusions cannot be extended to all built views, nor to other patient groups, such as long-term patients, who may suffer from low arousal or boredom rather than the anxiety problems associated with surgeries. Perhaps to a chronically understimulated patient, a built view such as a lively city street might be more stimulating and hence more therapeutic than many natural views. These cautions

notwithstanding, the results imply that hospital design and siting decisions should consider the quality of patient window views.'

This understated and thoughtful piece of research, the main aim of which seems to be to argue for improved hospital design, has become the standard evidence for the link between the natural environment and healing. Professor Ulrich is a world authority on hospital design and is an advocate of the provision of nature in a clinical setting. However, some of his research counters this – subsequent research at Uppsala University Hospital in Sweden in 1993 used pictures of landscapes rather than views.[41]

The use and enjoyment of open spaces is complex, with cultural and social dimensions that make it difficult to compare experiences or demonstrate a universal benefit. The experience of visiting a park is different for each visitor. The view that green space has a beneficial health effect is supported by research findings but it is hard to establish a causal relationship.

A 2011 review of the evidence of health benefits of urban green spaces published in the *Journal of Public Health* concluded that 'There is weak evidence for the links between physical, mental health and well-being, and urban green space. Environmental factors such as the quality and accessibility of green space affects its use for physical activity. User determinants, such as age, gender, ethnicity and the perception of safety, are also important. However, many studies were limited by poor study design, failure to exclude confounding, bias or reverse causality and weak statistical associations.'[42]

For example, one study reported a positive association between lower stroke mortality and higher

ULRICH'S VIEW
THROUGH A WINDOW

Records on recovery after cholecystectomy (gall bladder removal) of patients in a suburban Pennsylvania hospital between 1972 and 1981 were examined to determine whether assignment to a room with a window view of a natural setting might have restorative influences. Twenty-three surgical patients assigned to rooms with windows looking out on a natural scene had shorter postoperative hospital stays, received fewer negative evaluative comments in nurses' notes, and took fewer potent analgesics than twenty-three matched patients in similar rooms with windows facing a brick building wall.

levels of greenness in the environment. Whilst there is strong evidence of health benefits of physical activity the evidence for the link between physical activity levels and green space availability is less clear – people are more likely to use green spaces if they are available, but the open space provides the opportunity for exercise rather being the cause of the health benefit.

The 2016 World Health Organization (WHO) report 'Urban Green Spaces and Health – A review of evidence' emphasises the role of green space and the opportunity for exercise and play. The importance of green space provision has been recognised by the WHO European Region who have made a commitment '...to provide each child by 2020 with access to healthy and safe environments and settings of daily life in which they can walk and cycle to kindergartens and schools, and to green spaces in which to play and undertake physical activity'.[43]

As part of the 17 Sustainable Development Goals set by the United Nations in 2015 and adopted by more than 150 world leaders, they set the goal that 'By 2030, provide universal access to safe, inclusive and accessible, green and public spaces, in particular for women and children, older persons and persons with disabilities'.[44]

URBAN GREEN
SPACES AND HEALTH

'Urban green spaces, as part of a wider environmental context, have the potential to help address problems "upstream", in a preventative way – considered a more efficient approach than simply dealing with the "downstream" consequences of ill health.'[45]
Urban Green Spaces and Health report, 2016

We do need to explain the benefits of our work but there is a risk that if we rely on unfounded research, and see the provision of open space simply as a service with an economic benefit, the same reasoning can be used against us if that benefit can be delivered in another way. This isn't an argument against the provision of open space or the benefits of spending time outdoors – quite the opposite.

This point was made by Professor Paul Ekins OBE in 2009 at a conference exploring ecosystem services and human wellbeing.[46] Professor Ekins, Professor of Resources and Environmental Policy at University College London, warned that environment professionals should look to other sectors such as law and medicine for examples of how to justify providing health benefits or retaining habitats. He argued that some issues override financial value, such as human rights, justice and democracy. The legal sector takes a moral rather than financial perspective – legal cases are pursued because society feels it is morally right to do so, not because of the money saved or recovered. His concern was that if we put too great an emphasis on the economic case we run the risk of our arguments being turned against us if a function can be provided by less costly but more environmentally damaging means. If we only argue on an economic basis there is always the risk that there could be a trade off.

Fig 4.13 Complex paving patterns are stunning but can be challenging for those with a neurological condition, 2018, UK

Sensory perception

Experiencing a landscape is a sensory experience. Whilst many sensory experiences are pleasurable and are something that many people seek out, such as the scent of a flower or the sound of running water, each person has a different reaction to sensory stimulus. For some users what would be seen by most people as pleasurable is an uncomfortable or even unbearable experience.

In our work as designers we need to manage the impact of that sensory experience, where possible giving users a choice of what they experience, giving them agency over their environment. Understanding the potential negative impacts for some users can help us improve the spaces we design – many users may experience a negative response but on a level that is less perceptible, so they will also gain from a more sensory friendly scheme. What works for neuroatypical users, such as those with autism, also works for neurotypical users.

Senses

As children we are taught the five basic senses of sight, sound, touch, taste and smell but there are other hidden senses that are important for the designer to acknowledge. People with differences in how they sense the world may find our sites difficult to navigate or uncomfortable to visit.

Vestibular – The sensory system that responds to the position of the head in relation to gravity and movement. It gives us awareness of our position in the environment. This sense informs us of motion and our current orientation.

Proprioception – The sense of where our body is in space and the perception of the individual movement of body parts. It allows us to plan our movements without the use of sight or touch. This sense can be affected by age, disease and health issues such as arthritis, brain injury and poor circulation.
Together these two senses give us awareness of movement, speed and pressure on our joints and muscles.

Interoception – The sense of the internal state of the body, using sensory signals.

Deep pressure – The sensation of a hug, a child being held in the arms, a cosy chair or a weighted blanket. This pressure acts as a calming or focusing agent, increasing serotonin and dopamine levels in the brain.

All humans need a sensory diet – as we are evolved to be alert to danger a total lack of any stimulus can be unsettling, as can too great a level of stimulus. Depending on our mood we can be sensory seeking or want to avoid our senses, and we are happiest when we are in control of how much sensory stimulus we receive. Overstimulation of the senses can cause anxiety, restlessness and chronic stress.

Different people favour different types and levels of sensory stimulus – a loud thrash metal concert might be exhilarating for some but cause anxiety in others. Neurological conditions such as autism and Alzheimer's disease can make people more sensitive to certain sounds or lighting conditions, in some cases causing severe distress in a setting others would perhaps not even acknowledge.

Unusual textures, strong linear patterns in paving or high-shine surfaces can all cause problems for those with atypical sensory function, with changes in pattern being perceived as changes in level or high-shine surfaces perceived as water. Complex patterns can also cause visual disturbance in those with a sensitivity. In contrast natural patterns, such as trees moving in the wind or the movement of water, are seen by the brain as a neutral frequency and provide what Steve Maslin describes as 'visual bathing'.

Chad Kennedy, a landscape architect at California-based multi-disciplinary practice O'Dell Engineering, suggests the following concepts when designing to support those with a sensory disorder, especially recreational activities:[47]

Vestibular

- Provide wayfinding materials within the landscape
- Provide clear, precise visual cues related to circulation and activities
- Provide physical activities with varying levels of required motion (activities requiring ranges of intense movement to minimal movement)
- Provide activities and equipment that limit movement of the head
- Provide rest areas in close proximity to all active areas (a retreat from sensory-rich environments)
- Provide opportunities for postural security (this may include physical supports such as chairs,

tables, and vertical posts. Also included are the inclusion of strong vertical cues or impressions to help with upright orientation and the creation of cozy spots where individuals can escape from the immediate environment)

Proprioceptive

- Provide simple equipment with simple tasks
- Provide clear visual markers in the landscape
- Provide moveable items of varying weights to accommodate users of differing abilities and to promote muscle building activities
- Provide opportunities for stretching and compression activity (gymnastic bars, trampolines, climbing/tumbling, etc.)
- Include swimming pools (water pressure and buoyancy adds to sensory awareness)
- Provide rest areas in close proximity to all active areas (a retreat from sensory-rich areas)
- Provide physical supports such as chairs, tables, posts and rails to lean against, sit upon and re-gain stability

Understanding that different users will interpret our work in different ways can help us understand the impact of our work. Looking at how we can provide each type of stimulus but also where we need to create areas of low stimulus, such as common spaces used by all, will improve the quality of our designs and make the space more usable and comfortable to experience.

Old friends hypothesis

Recent research by Professor Graham Rook, Emeritus Professor of Medical Microbiology at University College London, has shown that the benefits of access to open space might be at a microbial as well as a physiological or psychological level.[48] His work suggests that the long-term health benefits of living close to the natural environment, such as reduced death rates, reduced cardiovascular disease and reduced psychiatric problems, are linked to exposure to the microbes that have evolved in parallel with humans.

Professor Rook acknowledges that there may be psychological benefits from accessing open space, perhaps evolved from 'habitat selection' which results in psychological rewards from approaching ideal hunter-gatherer habitats, as well as secondary benefits from social interactions, exercise and sunlight. However, he argues that much of the research demonstrating the value of open space fails to include comparable controls – would meeting friends in a favourite café or watching a feel-good film in an urban cinema have the same psychological effect? The research also fails to explain how exposure to open spaces provides long-term benefits to health.

Professor Rook believes that there is an immunology component that runs in parallel to the psychological component. To become effective our immune systems need a variety of microbial exposures to learn what is and what isn't a threat, and to prevent the body from attacking itself, harmless allergens or the gut. Many of the diseases of higher-income countries, such as atopic asthma, cardiovascular disease and depression, are linked to this misplaced immune response.

The microbes come from numerous sources, including animals, other humans, parasitic worms, water, plants and soil. A diverse range of microbia helps maintain a healthy immune system, so areas with a low diversity of microbes, such as areas of agriculture mono-culture or urban areas, could have an adverse impact on health. Also, lack of interaction with other humans can reduce exposure, compounding the negative health impact of social isolation.

This concept helps explain anomalies, such as the rise of immune-related illness in high-income countries as health care improves but access to the natural environment decreases, and how living in a greener environment can benefit health even if we rarely interact with the outdoors. Professor Rook draws a link between biodiversity of the natural environment and the regulation of the immune system.

This theory is a development of the hygiene hypothesis, where our genetic evolution hasn't caught up with cultural and technological evolution, and that while advances in the understanding of hygiene have helped saved millions of lives they have also reduced our exposure to organisms that we would have experienced as hunter-gatherers. We have an evolved dependence on these organisms and their absence has a direct impact on our health.

This research may help us further explain the benefits of urban green space, and he argues that 'If a

significant part of the role of the natural environment is to provide an appropriate airborne microbiota, then multiple, small, widely distributed urban green spaces of high microbial quality might suffice as supplements to a core of large recreational parks. There is already huge interest in the construction of roof gardens, vertical gardens and urban green spaces motivated by aesthetics and by organizations wishing to promote urban habitats for birds and insects and by urban planners wishing to delay the entry of rain downpours into sewer systems. However, we suggest that combating the epidemic of inflammation-associated illnesses in high income urban environments provides another compelling motive for creating green spaces, and we hope that this paper will enhance collaboration between the medical profession, ecologists, and urban planners.'

As landscape architects we should recognise Professor Rook's suggestion that microbial diversity should be acknowledged as an ecosystem service, and that further research is needed to refine our understanding of the process. It will be interesting to see if our future will include maximising the microbial diversity and output of our schemes, or designing schemes that have no public access but still have health benefits. Perhaps the incidental spaces left between urban developments, such as the area under billboards, dead-end alley ways or small areas of unused public land could be used for this purpose.

Our understanding of the role of open space and natural environments is still at an early stage. Throughout history many cultures have seen access to outdoor space as an essential, perhaps as a place

Fig 4.14 (Page 126) Village green – even in rural areas the provision of public open space is important, 2016, Grasmere, Lake District National Park

for religious contemplation such as a cloister, or as a means to better health, such as the cure walks taken in the countryside by those seeking improved health in Georgian Bath. The rationale behind the benefits of access to the natural environment may change but the consensus of these benefits has outlasted each theory.

Our role should be to argue the point that all adults and children should have the option to visit safe, accessible open spaces close to their homes and that open spaces shouldn't be expected to single-handedly solve social issues that should be addressed elsewhere, such as improving physical and mental health or addressing poor air quality.

Provision of open space should be viewed as a moral requirement as opposed to an economic function. Many of the benefits are subtle and hard to measure, such as the enjoyment of beauty or the opportunity to be with other people. Public open space provides the opportunity to be in a setting other than your own home or garden, to sit in the shade on a hot day or to find breathing space away from home responsibilities.

The client commissions our work but we have a wider responsibility to understand and manage the implications of our designs. The implications can be far reaching and far outlive us. Whatever the budget, timescale or other limitations on our work as design professionals we cannot, and should not, absolve this responsibility.

'Design is a process of creating a solution to a brief and then preparing instructions allowing that solution to be constructed.'
Design Coordination, Designing Buildings Wiki www. designingbuildings.co.uk

The design stage may be the stage that our training focuses on most heavily but in practice we may not have as much time to give to it as we'd like. Whatever the time or constraints our job is still to understand the client's needs, create a design that meets those needs and then communicate that design to others. However involved we get in the process and however creative the solutions we come up with, the scheme always belongs to the client.

The design stage can be enjoyable, frustrating and rewarding in equal measure. We need to remember that we are meeting the client's needs, not our own, and even if we dislike the solution if it meets the client's requirements and will function as a viable landscape scheme, we have done our job. We have taken the germ of an idea and developed it into a scheme that can be built by others. Elements and dimensions and have been agreed. The project has moved from our imagination into almost being a reality.

Case Study 4.1
WHITEHALL, LONDON
TERRORISM AND SECURITY

Title
Whitehall Streetscape Project

Client
Cabinet Office

Location
Great George Street, Horse Guards Avenue,
Horse Guards Road, Parliament Square, Whitehall,
Whitehall Court, Whitehall Place, London, UK

Design period	Construction Period
2003–2007	2007 to 2010t

Type of Scheme	Project Value
Public realm	£30 million

Owner
Westminster City Council

Project Team

Design team and contractor - West One Infrastructure Services (consortia of Hyder Consulting, JM Murphy & Sons Ltd and FM Conway Ltd)

Security engineering consultants - MFD Group

Archaeology Pre-Construct - Archaeology Ltd (London)

Architectural consultants - Purcell Miller Tritton LLP

Public realm consultants - Atkins Streetscape

Traffic modelling - Colin Buchanan and Partners (now SKM Colin Buchanan)

Security and hostile vehicle mitigation consultants - MFD Group and TRL)

Suppliers

Specialist steel fabrication - Corus Group plc (now Tata Steel) using Corus Bi-Steel

Masonry contractor - Stone Restoration Services Ltd

Fig 4.1.0 (Page 129) Whitehall at night with Union Flag pennants, 2011, Whitehall, London

Fig 4.1.1 (Page 130) Whitehall is a busy thoroughfare also used for state occasions – the Cenotaph is the UK's official war memorial and the focus of the annual Remembrance Service, 2018, Whitehall, London

Attacks in London, Manchester, Nice, Stockholm, Charlottesville and other cities, where vehicles were used to deliberately ram buildings, crowds or other vehicles, show that there is a small but very real risk to the public when they use the spaces we design.

As a consequence structures to prevent vehicle attacks have been installed in areas seen as high risk. Done well these important structures can add to the street scene, allowing an opportunity for art installations, new seating or play structures that subtly act as defensive structures without drawing attention to the potential risk. Done badly security measures can sit uncomfortably in the landscape, changing the character of a space.

Whitehall, the world-famous street in the heart of London, is the location of many government buildings including the Cabinet Office, the Ministry of Defence and the Foreign Office, as well as being the home of the Cenotaph War Memorial and the access point to Downing Street. The nature of the buildings means that security is a major issue but it is also a busy road, with over 18,000 vehicles passing through each day.[49] Running from Trafalgar Square to Parliament Square, it is also a popular tourist route.

The scheme area included 78 listed buildings, 14 of which are Grade 1 Listed, and the street is at the centre of the Whitehall Conservation Area.[50] Part of the site was within the Victoria Embankment Gardens and St James's Park lies to the west. Both of these are designated Parks and Gardens of Special Interest. The UNESCO World Heritage Site of the Palace of Westminster and Westminster Abbey including St Margaret's Church abuts the south edge of the site.

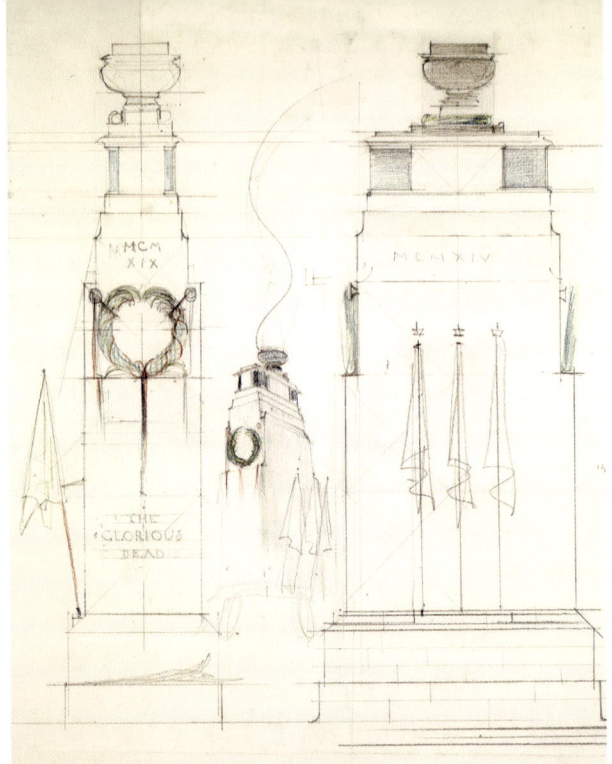

Fig 4.1.2 Design for the Cenotaph, Whitehall, London, surmounted by an urn: sketched elevations and perspective, Sir Edwin Lutyens, 1919

'You must no more call it York Place – that is past: For since the Cardinal fell that title's lost; 'Tis now the King's, and called Whitehall.'
Shakespeare's Henry VIII Act IV, sc. 1

Evidence of activity during the Mesolithic, Neolithic and Bronze Age periods has been found in the area, along with Roman and Saxon items.

Security is not a new issue for Whitehall – in the 17th century the original Palace of Whitehall was protected from attack by posts around the approach to the gateway. Public and private buildings have often included such security features, with earlier buildings including layers of defence such as the motte and bailey or moats used to weaken any attack and to limit the points of entry.[51]

Government bodies on Whitehall had wanted to improve the security from vehicle borne terrorist attacks including those from vehicle bombs which had manifested elsewhere.'[52]

This provided an opportunity, and funding, to improve the street for residents, businesses and visitors as well as including new security measures.

The brief for the project was to deliver two distinct and separate requirements – the work coordinated by the Cabinet Office to improve the security for Central Government departments and Westminster City Council's wish to transform the area into a safer, better environment for all.

Fig 4.1.3 Improved paving in keeping with the historic setting, 2018, Whitehall, London

Fig 4.1.4 Balustrade that incorporates a vehicle security barrier, 2018, Whitehall, London

Fig 4.1.5 New walls and bollards allow pedestrian movement whilst providing protection from vehicle-borne attack, 2018, Whitehall, London

A direct route towards an asset allows a hostile vehicle to build up speed on approach.

Chicanes and offset approaches to an asset reduce hostile vehicle approach speed.

Moving a road, or an asset, to create an indirect approach will lead a hostile vehicle away from the asset.

Removing vehicle access from the front of an asset removes the potential for using a vehicle as a weapon and establishes a stand-off distance from parked hostile vehicles.

The decision was taken to work collaboratively – 11 different government departments pooled their resources under Cabinet Office lead, in turn contracting with Westminster City Council and their infrastructure services provider West One.

The project team consulted with over 50 organisations including central and local government, and local interest groups. Creating a design that met the complicated and conflicting functional needs of multiple stakeholders, satisfied the statutory requirements of organisations including Transport for London and English Heritage, incorporated enhanced anti-terrorist barrier systems and could be constructed with minimal disruption to a busy route in a Conservation Area adjacent to a World Heritage Site was a challenge for the design team.

The new scheme included:

- installation of security bollards and architecturally sensitive balustrade walls
- footways re-laid and widened to improve pedestrian access
- carriageway reduced to two lanes in each direction
- reduction in street clutter
- upgraded lighting
- redesign of the walls and gardens around the Ministry of Defence Main Building
- improved paving.

Fig 4.1.7 Retained mature trees, 2018, Whitehall, London

Fig 4.1.6 (Page 134) The scheme uses the principles set out in *A Public Realm Design Guide for Hostile Vehicle Mitigation – Second Edition*[53]

The scheme was designed with defence and protection as a primary function but this is not reflected in the character of the design – the wider footpaths and well-laid-out crossings make it a welcoming space for visitors. There are no overt hostile design features, such as those mentioned earlier in the chapter. The street does have a high level of human surveillance to deter low-level anti-social behaviour, with armed police at locations such as the entrance to Downing Street, but it is also designed to be accessible and inclusive.

Further information

- The Centre for the Protection of National Infrastructure https://www.cpni.gov.uk/hostile-vehicle-mitigation
- Vehicle Security Barriers within the Streetscape https://www.gov.uk/government/publications/vehicle-security-barriers-within-the-streetscape
- Integrated Security – A Public Realm Design Guide for Hostile Vehicle Mitigation – Second Edition https://www.cpni.gov.uk/hostile-vehicle-mitigation (file named Integrated Security)
- Resilient Design Toolkit – Counter Terrorism http://www.securedbydesign.com/industry-advice-and-guides/

Fig 4.1.8 Vehicle-deterrent bollards with Horse Guards in background, 2018, Whitehall, London

Fig 4.1.9 Protective wall outside Ministry of Defence Main Building on the site of the Palace of Whitehall, designed to complement the style of the Emmanuel Vincent Harris classical building, built 1938–1951, 2018, Whitehall, London

Case Study 4.2
MERTON BORDERS
PLANTING FOR A DRIER, HOTTER FUTURE

Title
Merton Borders

Client
Tom Price – Gardens Curator,
University of Oxford Botanic Garden

Location
University of Oxford Botanic Garden, Oxford, UK

Design Period	Construction Period
2005	2011
Type of Scheme	**Project Value**
Botanic garden	unknown

Landscape Architect
Professor James Hitchmough, Professor of Horticultural
Ecology at the University of Sheffield

Owner
University of Oxford Botanic Garden

Suppliers

Seed supplier - Jelitto Perennial Seeds, (Jelitto
Staudensamen GmbH), Germany

Jute matting - Soil Saver™ biodegradable erosion control
mesh, Hy-tex, (UK) Limited

In 2008 the University of Oxford Botanic Garden
commissioned Professor James Hitchmough, Professor
of Horticultural Ecology at the University of Sheffield,
to design a new area of planting. The area, which was
first planted in the 1940s, had become over-mature so the
decision was made to clear the entire area and create a
new scheme. The brief was for a naturalistic sustainable
planting scheme that would have minimal long-term
impact on the environment, be drought tolerant with
no irrigation after establishment, require no staking,
soil improvement or fertiliser and require no intensive
maintenance techniques such as division or replanting.

The solution was to create highly ornamental yet
sustainable plant communities based on drought-
tolerant seasonally dry grassland communities in three
regions of the world.

Fig 4.2.2 Structure in the first growing season is provided by
material planted at a density of approximately 1 plant/2-3m².
Many small sown seedlings can be seen that will fill in the spaces
by 2013, Oxford Botanic Garden 2012, Oxford

Fig 4.2.0 (Page 137) Echinacea in American prairie community;
Oxford Botanic Garden June 2014, Oxford

Fig 4.2.1 (Page 138) American prairie community; Oxford Botanic
Garden June 2014, Oxford

The regions are:

- The Central to Southern Great Plains (USA) through to the Colorado Plateau and into California
- East South Africa at latitudes above 1000m
- Southern Europe to Turkey and across Asia to Siberia

The biogeographic regions are all 40 to 30° from the Equator, 10° closer than Oxford. This is based on climate change modelling that predicts the future climate of south-eastern England will resemble that of present-day Bordeaux.

To avoid using plants grown in plastic pots in a peat-based compost 85% of the plants were direct sown. Sowing seed rather than planting container-grown plants allowed for far greater planting densities, in turn achieving a higher level of diversity and a longer succession of flowering period. The seed mixes were prepared to a precise recipe and mixed with sawdust to sow onto a raked and rolled 75mm sand mulch. The mulch was principally used to reduce weed seed germination but it had the secondary benefit of increasing the reliability and robustness of many of the drier climate species.

The areas were raked again after sowing and a loose-weave biodegradable jute net laid to hold together the sand and seeds, create a microclimate to encourage plant development and to prevent digging by animals. The area was irrigated for the first 12 weeks. After the initial sowing in spring 2011 the area was bare. The area was over-seeded in autumn 2011 and in the summer of 2012 weeds were still present. However, by the 2013 the area was fully established and had become a recognised attraction within the Botanic Garden.

Fig 4.2.3 Over-sowing previously planted material, Oxford Botanic Garden, November 2011, Oxford

Fig 4.2.4 Spring growth, Oxford Botanic Garden, May 2015, Oxford

These species are the antithesis of the tall, leafy herb plants dominant in UK planting design, that favour moist to wet ground conditions that few sites can sustainably provide over our increasingly dry summers. The sand mulch helped reduce the soil fertility, making it closer to the levels found in natural systems.

Fig 4.2.5 Early summer growth in South African community showing sand layer, Oxford Botanic Garden, June 2014, Oxford

Fig 4.2.6 *Stipa gigantea* continues to provide structure into autumn in the Eurasian community, Oxford Botanic Garden 2016, Oxford

In enriched soils the volume of foliage is greater, in turn creating more competition for light, favouring the larger, vigorous species.

The layout uses strong, geometric shard shapes sown by region, with each region repeating across the site. Some plants are repeated across all the areas as an overlay. Nearly all the plants have basal foliage and a naked flowering stem – this reduces shading. The result using this plant form is three distinct layers – a low, sheet layer, a bumpy layer and an emergent layer. The design doesn't include the traditional tiering of many planting schemes with lower plants towards the edge. As a result of direct sowing the layout has a random, open character with plants repeating all the way through each zone area. This complex, layered vegetation contains few UK native species but still provides a valuable invertebrate habitat.

The plant communities in the American and South African zones are tolerant of being burnt in the spring, so the management of these areas includes a flash burn to kill off small weeds without damaging the planting.

Direct-sown planting has the advantages of being cheaper than using pre-grown plants, and allows tightly packed plant communities to evolve to suit the site conditions and resist weeds, and it can be a good way to make use of sites with low fertility. The establishment phase does require more input that other planting techniques, with the potential for substantial weeding and a stark appearance in the early stages but it is a low-input, sustainable approach that incorporates climate resilience.

Fig 4.2.7 (Pages 142 and 143) Summer view of the Eurasian community, Oxford Botanic Garden, August 2016, Oxford

Fig 4.2.8 (Page 143) South African community, Oxford Botanic Garden 2015, Oxford

Fig 4.2.9 (Page 143) Long view through the three plant communities, Oxford Botanic Garden July, 2013, Oxford

Chapter Five

CONSTRUCT
AND MANAGE

INTRODUCTION

This is the stage where your project moves from a concept into reality. Dimensions and layouts are set out, allowing the client and the rest of the project team to see the design in situ. New issues come to light as the project develops, and with staff on site there is a greater urgency to solve these quickly to prevent delay.

As designers we now need to take a leap of faith and pass our ideas to others to interpret. We are moving closer to meeting the client's needs, identified during the planning stages. Changes during previous stages may have had cost or time implications but there was no physical impact of a redesign. Once the first sod is cut changes can have complex and wide-ranging impacts, especially if items have been ordered. Ensuring that the client and the rest of the project team understand the potential impact of changes at this stage, and that all possible decisions have been made before construction begins, is an important check. Rushing into the construction phase with too many questions left unanswered, perhaps driven by a fixed deadline, is rarely a good strategy. Time spent planning and resolving issues is normally recouped.

Much of the good practice from earlier stages continues – clear communication, collaborative working and accurate management of information are still central to successful working. New issues at this stage include the risk of the unexpected, such as unforeseen discoveries on site, poor weather conditions or external factors such as a problem with the supply

of materials. Soil sampling may have missed spots of contamination, or archaeological remains could be revealed. The pre-mortem technique mentioned in Chapter 4 can help identify potential risks but systems need to be in place, such as appropriate contingency levels and quick decision-making processes, to ensure that changes can be managed effectively. Assuming that no changes will occur is naive, and all members of the project team should be aware of the potential for unforeseen change.

Depending on the form of contract this is likely to be the point the contractor joins the project, along with suppliers and specialist sub-contractors. With more individuals involved and complex, often expensive construction processes about to start, efficient systems need to be in place. Setting up processes to manage this

Fig 5.1 Work to recreate historic drainage grips on the Gwent Levels – on historic sites the risk of unearthing archaeological finds is high; 2019 Gwent Levels, Wales

Fig 5.0 (Pages 144 and 145) Bluebells, 2012, Oxfordshire

STAGE 5 – CONSTRUCTION

- Manage implementation of handover strategy
- Project team meeting
- Review and update project execution plan
- Pre-start meeting
- Comment on construction programme
- Site progress meetings
- Monitor and review progress and performance of project team
- Carry out site inspections and review
- Respond to site queries related to coordination or integration
- Respond to design queries
- Update construction strategy
- Coordinate site inspections
- Monitor construction progress
- Check design sustainability assessment
- Check sustainability procedures
- Review handover strategy
- Check non-technical user guide
- Design team meeting
- Carry out design/technical review
- Undertake final site inspection
- Prepare defects report
- Prepare 'as constructed' information
- Check 'as constructed' information
- Exchange 'as constructed' information

STAGE 6 – HANDOVER AND CLOSE OUT

- Manage tasks listed in handover strategy
- Manage updating of 'as constructed' information
- Project team meeting
- Monitor and review progress and performance of project team
- Undertake tasks listed in handover strategy
- Review protection of protected habitats and species (and other statutory protections such as TPO) and invasive species
- Design team meeting
- Carry out design/technical review
- Update 'as constructed' information
- Review updated 'as constructed' information
- Manage preparation and issue of 'as constructed' information by specialist sub-contractors
- Advise on the resolution of defects
- Conclude administration of landscape contract
- Review project information
- Check 'as constructed' information
- Exchange updated 'as constructed' information
- Undertake end of defects inspection
- Hand over health and safety file
- Completion and establishment
- Contract preparation for landscape maintenance

Table 5.1 The elements of this workstage[1]

BIM LEVEL 2 BENEFITS MEASUREMENT

A 2018 report by PricewaterhouseCoopers LLP (PwC) commissioned by Innovate UK gives an interesting insight into the benefits of using BIM Level 2 (BIML2).[2] The report uses two public-sector case studies – the refurbishment of the Department of Health headquarters at 39 Victoria Street in London and the Foss Flood Barrier in York – to consider the benefits for using BIM Level 2 when working on public-sector assets.

The report looks at the whole life cycle of each example, taken as approximately 12 years for Victoria Street and approximately 24 years for the Foss barrier, so much of the saving relates to the operational stage. However, there were measurable savings during construction attributed to the use of BIM.

Table 5.2 Percentage benefits estimated in each phase of life cycle

PROJECT	DESIGN STAGE	BUILDING AND COMMISSION + HANDOVER	OPERATION
39 Victoria Street	6% £42,366	21% BC £103,872 Handover £84,520	73% (~12 years) £391,592
Foss Barrier Upgrade	36% £132,217	3%* BC £5,757 Clash detection £6,500	61% (~24 years) £223,118 Equivalent to 6–7% per annum
Other benefits identified by stakeholders but not quantified include (both projects)	– Time savings in design coordination and management	– Time saving in construction schedule planning and quality control – Cost savings from clash detection and fewer changes – Environmental benefit from fewer materials used	– Improved asset quality – Health and safety benefits in maintenance – Improved reputation – Time savings in incident response

* BIML2 was not a key part of the review process, as there was a small team all working on site. The design stage and the building and commission stage ran concurrently so these savings are allocated to the design stage.

Fig 5.2 Identifying poor-quality work is central to our role – fencing work not undertaken to specification, 2007, Berkshire

THE LANDSCAPE ARCHITECT

Our place in the project team determines our experience of the construction phase. If we are lead or sole consultant we will be central to the process, working directly with the client and heavily involved in the day-to-day delivery of the project. If we were a sub-consultant during the design stages our role may be minimal, perhaps only becoming involved if there are changes or a problem comes to light. If we are working as part of a building or infrastructure project with long timescales we might only become involved as the work we designed is implemented, hoping that the landscape budget survives intact, or we may be involved throughout ensuring that tree protection, soil handling and other landscape issues are dealt with correctly.

As the project is constructed our role changes from envisaging the scheme to ensuring that what is built is as close as possible to the designed scheme, taking into account constraints such as site conditions, budget and timescales. Depending on our level of involvement we might undertake regular site inspections – we need to make sure that we are confident that we can identify poor quality work that has not been completed to the agreed specification. For complex processes, or features that need to tie in with existing features, sample sections should be agreed, and this used as a benchmark for the remaining area.

However thorough the design process has been there are likely to be changes during the construction phase, as new issues become known or costs are finalised. Managing this change, and helping the client understand their choices, and the implications of

stage, including how changes are managed, powers of instruction and site inspections, will pay dividends as the project progresses.

Roles need to be defined and risks managed on a constant basis. If we are leading the project team we need to ensure that roles are allocated and each team member knows what level of information others require. If we are part of the project team we need to be clear about our own role within the team, and that other team members understand the scope of our work. Collaboration, an ethos that should have started during the design stages, is now even more important.

each change, are central to our role at this stage. The life span of the landscape elements within a project, and the fact that we deal with living materials that change over time, means that the implications might not be immediately clear, and the impact might not be felt for decades, but we still need to highlight the potential impact. However large the trees are that we plant they are usually a fraction of their final height and width. Remembering that a landscape scheme is dynamic, and will constantly change, is a point worth reinforcing.

If we are leading the project we need to monitor progress and update the client, advising on the impact of any delays and identifying any opportunities to mitigate.

Landscape architect as lead consultant

The landscape architect typically acts as lead consultant if the landscape components are the main or only works. In other scenarios, such as projects with a substantial built element, it is less common for the landscape architect to act as lead, although there are examples where clients have appointed the role to the landscape architect.

In situations where buildings, facilities and routes need to be defined within a landscape setting, perhaps following the design stages from Hal Moggridge's book discussed in Chapter 2, it would make sense for the landscape architect to act as lead consultant, as the initial steps are landscape-led. The long establishment period of many landscape elements means that our involvement is often over years rather than months.

Fig 5.3 London 2012 Olympic Park – one of the largest UK landscape schemes in recent times and an excellent example of built elements integrated into a complex landscape, 2012, Stratford, London

Unlike many of the other professionals on the project team we have an interest in the entire site, right up to the boundary, as well as considering the wider landscape context. All these factors mean that there is often a strong argument for the landscape architect to act as lead consultant.

Collaboration with other landscape practices

When setting up a business, potential business owners are encouraged to research their competition to help them understand the market and to decide whether there is a need for their service. Instead of seeing competitors as the enemy in my experience other landscape architects who are technically competitors are also potential collaborators, and a wonderful source of support.

For smaller practices this collaboration can help with fluctuations in workload by sub-contracting tasks, or by providing additional expertise where specialist knowledge is required. Larger projects might be beyond the capabilities of one practice, but by partnering with other landscape architects who share your values you can bid for new areas of work. Knowing, and trusting, other landscape practices also means that you can refer job enquiries that aren't suitable for you to someone you know is reliable, keeping your reputation intact.

With so many landscape architects employed in small practices, often with only one office, there is also scope for practices to help each other with minor tasks. Spending a long day on site, only to discover you missed one small but important detail, can be frustrating but also costly if you need to travel a long way back to site

to check just one item. As the error is yours the cost shouldn't be passed to the client. For complex tasks, such as photography for landscape and visual impact assessments or where a judgement needs to be made, you would need to return to site. However, for tasks with minimal consequences knowing a local landscape architect who can undertake the task for you can be a huge time saving.

Sharing the cost of training can be another way that smaller practices can collaborate. Again, it may feel counter-intuitive to help those who are your competitors, but sharing costs can make expensive training courses feasible, and by improving our skills we help improve the profession as a whole.

Issues such as client confidentially and professional liability would need to be considered, but having an informal network of other landscape architects you can turn to for help, something I have cultivated as I developed my practice, is a valuable resource.

Managing projects during the construction phase can be a difficult balance, with different risks depending on how we are appointed. If we are lead consultant we may be drawn into time-consuming issues that are not within the scope of our appointment but need resolving, making the project unprofitable if the time isn't covered by fees. If we are a sub-contractor the project may change substantially from our original design without our involvement, but still be credited to us, risking our reputation if the final scheme on site is not of a quality we would aspire to.

THE CLIENT

The construction stage is where the balance of risk moves even further towards the client – up to this point the costs may have been substantial but nothing has been constructed, so changes have been simple.

Once construction begins the client is almost always committed to completing the works, unless circumstances change so dramatically that the project becomes untenable. The consequences of unrealistic budgets or timescales, either due to optimism or cynically to win and keep the project, mainly fall in the end to the client.

As construction begins an unrealistic budget can encourage the contractor to cut corners to maintain their profit margin, reducing the quality below the specification and disappointing the client. Contractual controls should mean that the client should never pay for sub-standard work, but if the budget is too low to achieve the specification this may be a consequence. Any cost savings can then be lost in disputes and corrections, making the unrealistic budget a false economy.

The client's costs move from paying professional fees to paying for work on site and purchasing of materials, increasing the payments by a significant order of magnitude. The client needs to understand the level of payment that will be required. For clients with complex funding, such as charitable grants or staged loans, cashflow may need careful planning to allow them to meet their financial obligations.

Construction is also the stage where the client can see that real progress has been made. After the intangible process of design, seeing the project begin to appear can be a reassurance that the project is coming to fruition. Clients sometimes push to start on site before all the required preparation work has been complete, perhaps setting up a site compound or clearing vegetation, as it is a milestone that can be reported to stakeholders and a visible sign that something is really happening. However, this can be a false economy as starting work before all issues are resolved can result in abortive costs.

The client's role during construction

The client usually commissions the project, sets the overall programme and pays for the work. If they are the final owner of the project, they have an interest in the maintenance and whole-life cost of the site as well as the construction costs. If they are not the ultimate owner of the site, perhaps acting as a developer, in use costs or usability may not be as important. However, reputation is important to future sales, so a site that doesn't meet the needs of the ultimate user can impact on the success of the client's subsequent projects.

The client sets the tone for the whole project – if they are conscientious about health and safety, and state it as a priority, then it is likely that the rest of the project will follow their example. The additional tasks the client takes on, over and above those listed in Chapter 3, can include:
- publicity and media relations
- public engagement
- site safety
- permissions and consents
- managing site access

Fig 5.4 Temporary vehicle barriers in the city of Bath – now replaced by permanent defences more in keeping with the World Heritage Site, 2019, Bath

Communication with the client

As with the earlier workstages our contact with the client is decided by the arrangement of the project team. If we are lead consultant we will have direct access to the client, helping them manage changes and make the numerous decisions they will need to take as the project progresses. If we are a sub-consultant our contact with the client may be one or even two steps removed, with the risk of misinterpretation if others are making requests on our behalf.

The level of information that the client needs day to day should be agreed, especially during construction. A worried client is a risky client, as they can lose confidence and start changing their minds – they need enough information to be reassured but not so much that they are overwhelmed. An experienced client who works in our sector may want access to all project data so they can monitor progress in real time. An inexperienced client who is taking on the role over and above their normal work, such as a client running a school or managing a charity, may only want a weekly snapshot or only to be contacted if there is an issue to resolve. They might like to be updated by email, phone, instant messaging, updates from online collaboration tools or via a shared issue log saved to cloud storage. It is important that we manage the client's expectations, their available time and their preferred communication method.

Balancing publicity with security

The client is usually responsible for keeping the wider community updated on progress and notifying them of any disruption, with the support of the project team.

For controversial projects media coverage may need to be carefully managed. For some projects the full details can be shared publicly and real benefit can be gained by an open approach to engagement. The local community may not welcome every piece of work we undertake, especially if the construction process adversely impacts on their lives, but we can make any disruption easier to manage with good communication.

Other projects need a different approach, perhaps because of commercial confidentiality, or security issues. The work to protect Whitehall from vehicle attacks (see Case Study 4.1) is not a project where much detail could be shared. Similar work includes work on government embassies, defence sites, critical national

PAS 1192-5:2015 SPECIFICATION FOR SECURITY-MINDED BUILDING INFORMATION MODELLING, DIGITAL BUILT ENVIRONMENT AND SMART ASSET MANAGEMENT

The title of this PAS (Publicly Available Standard) may not seem immediately relevant to landscape architecture. BIM is associated for many with 3D modelling, rather than security. However, the openness and sharing of data that is central to BIM also brings the risk of that information being accessed. The standard gives guidance that should be considered on many of our projects, even if BIM is not being used. The guidance document 'Introduction to PAS 1192:2015' produced by the Centre for the Protection of National Infrastructure (CPNI) in partnership with the British Standards Institution (BSI) outlines the scope of the standard and the scenarios in which it should be used.[3]

'PAS 1192-5 specifies the processes which will assist organisations in identifying and implementing appropriate and proportionate measures to reduce the risk of loss or disclosure of information which could impact on the safety and security of:

- personnel and other occupants or users of the built asset and its services;
- the built asset itself;
- asset information; and/or
- the benefits the built asset exists to deliver.

Such processes can also be applied to protect against the loss, theft or disclosure of valuable commercial information and intellectual property'.

PAS 1192-5:2015 defines a sensitive built asset as 'one which, as a whole or in part, may be of interest to a threat agent for hostile, malicious, fraudulent and criminal behaviours or activities'. The response is always proportionate to the risk, but elements of our work may well need the level of scrutiny that the PAS requires. The standard also highlights that a security-minded approach may have business benefits even for projects not classed as high risk, such as reducing the risk of theft or data loss.

infrastructure or research centres linked to animal testing. Even on community projects where our default ethos is to be as open as possible, we may need to limit what is shared. Details of security measures, such as bollards to prevent vehicle attacks, or details of the areas covered by CCTV, shouldn't be disclosed, either during design or construction.

I have seen the installation of vehicle barriers in a city centre undertaken in full view of the public, revealing the structure and design of the barriers. This information could reveal any weakness in the design or even how to dismantle it. Making sure that security measures are screened from public view during construction is a small but important detail.

Progress and expectations

The process of constructing a landscape can be a disruptive experience. Removing vegetation and stripping topsoil can transform an innocuous site into an eyesore. For large sites the visual impact during construction may have been considered, and mitigation measures included to reduce the impact on the wider

Fig 5.5 Cleared site for screening bund and new ponds – the appearance of a cleared site can be a shock for clients; Thirlwall Associates for private client, 2008, Ledburn, Buckinghamshire

they now have to act to return the site to a usable state. An experienced client will understand this stage but an inexperienced client may need reassurance.

Planting schemes can be another area where the client's expectations need to be managed. As landscape architects we know that plants are often supplied as small specimens, but with the right conditions they soon grow and fill the space we have allowed in our planting plan. To a client the scheme can look sparse and underwhelming. Explaining the timescale for the planting scheme to develop, and showing examples of similar schemes and the rate of growth achieved, can help reassure them.

Smaller plant stock usually establishes better than larger stock when transplanted, is significantly cheaper and can have less environmental impact through reduced transport costs, less soil being removed from the nursery or minimal need for irrigation. Larger stock gives a more complete scheme from day one and can work well in sites such as public parks where immediate impact is needed, but can be expensive to buy and more complex to maintain.

If the client expects the instant landscape of a show garden or a TV show makeover they need to understand that this is expensive, and that the final planting densities may need to be reduced, with the associated waste of plant material. Once again the impact of time on a landscape scheme needs to be emphasised – any scheme is a balance between cost, impact and maturity.

The construction stage is the stage with the largest contrast, moving from concept to reality. It can be a difficult stage for an inexperienced client and in the day-to-day running of a busy site we need to make sure we find time to listen carefully to their concerns.

landscape, but however carefully site clearance is managed the change can be dramatic.

If the client knows a site well this can be a difficult stage, especially if the site has emotional connections for them, such as their private garden. Warning the client and other stakeholders at the outset that the appearance of the site is likely to get worse before is gets better is helpful, as it reduces the surprise, but it can still be a shock. Site clearance is also the point where the client often becomes fully committed to the project – regardless of the eventual scheme delivered

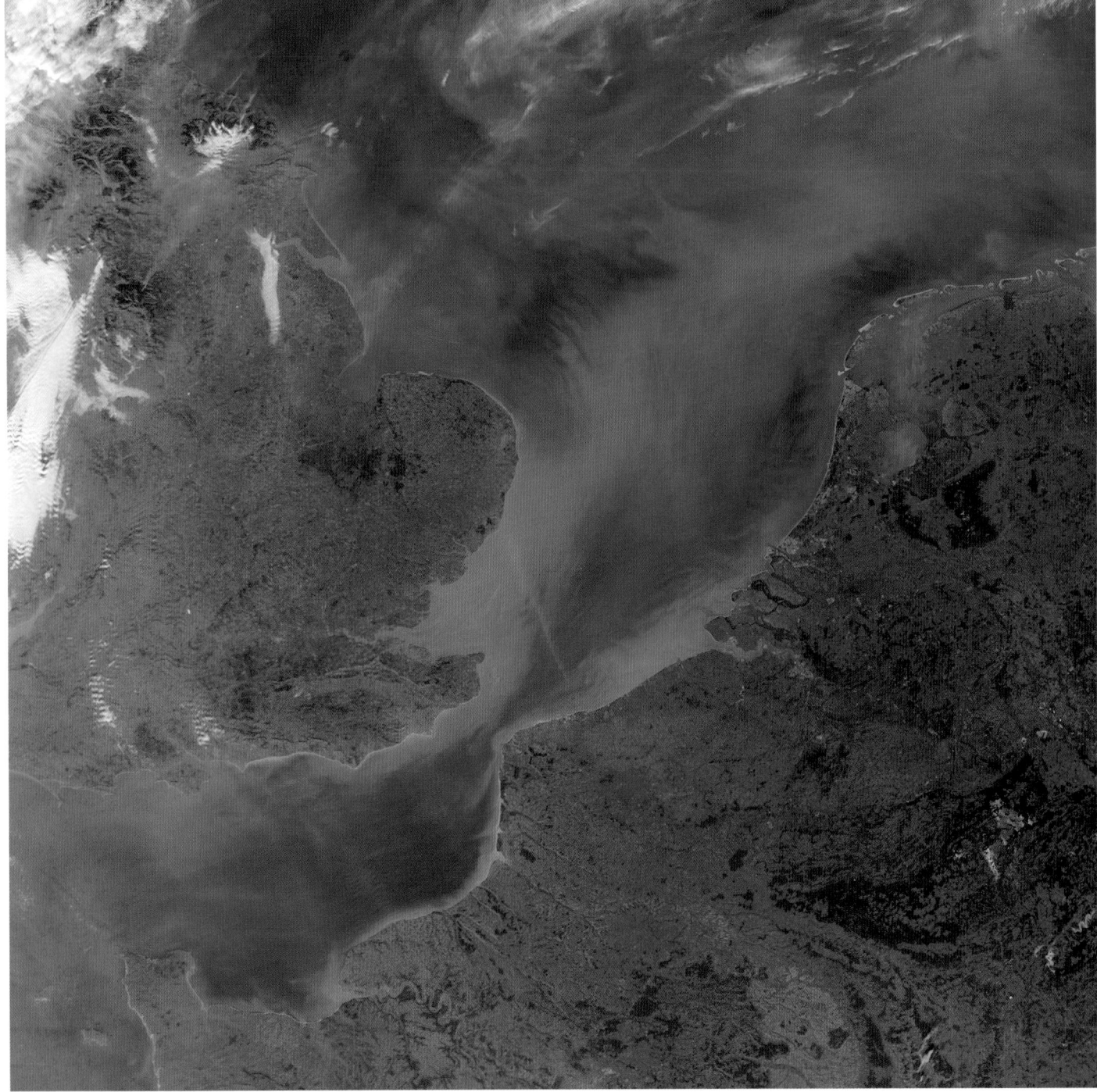

THE PROJECT

The project is now visible, and often subject to more scrutiny than during the design and planning stages.

Protecting what needs to be retained

As the project commences on site, the impact of the work needs to be managed. However comprehensive the site clearance it is likely that some elements will need to be retained.

Soil

In construction, soil – and in particular topsoil – is often overlooked as a resource. Soil is an ecosystem, made up of complex communities of micro-organisms including bacteria, algae and fungi and macro-organisms including earthworms and arthropods. It is a finite and essentially non-renewable resource – it can take up to 500 years to form a 2cm soil thickness.[4]

The 2009 UK government's Construction Code of Practice for the Sustainable Use of Soils on Construction Sites gives a useful list of the main functions of soil.

'Soil fulfils a number of functions and services for society which are central to social, economic and environmental sustainability. These are:
- food and fibre production;
- environmental interaction (with water and air);
- support of ecological habitats and biodiversity;
- support for the landscape;
- protection of cultural heritage;
- providing raw materials; and
- providing a platform for construction.

Soil also has a large social function, through providing the basis for green space, including gardens, playing fields and public open space. The latter provides cultural and social benefits that include increased well-being, physical and psychological health, and connection with nature. It consequently plays an important part in how people live.'

Unless the site topsoil is unusable, perhaps due to contamination, retaining and reusing the existing topsoil is usually the best option. The amount of topsoil needs to be assessed – even if the site is well vegetated it doesn't guarantee that the existing topsoil is of a suitable depth and quality for the planned landscape scheme. However, the existing soil can be improved if needed before being replaced. Stripping topsoil, moving it from site by vehicle, potentially paying to dispose of it then buying new topsoil to replace it is not a sustainable approach. Removing on-site soil that is unique to that site, that has taken hundreds of years to form, should be a last resort.

The Code of Practice lists the adverse impacts of the construction sector:
- covering soil with impermeable materials, effectively sealing it and resulting in significant detrimental impacts on soils' physical, chemical and biological properties, including drainage characteristics
- contaminating soil as a result of accidental spillage or the use of chemicals

Fig 5.6 (Page 156) Soil loss – clouds of brown and green sediment swirl through the North Sea, true-colour Aqua MODIS image acquired on 18 December 2004 (data date 18 December 2003); Jeff Schmaltz, MODIS Rapid Response Team, NASA/GSFC

- over-compacting soil through the use of heavy machinery or the storage of construction
- materials
- reducing soil quality, for example by mixing topsoil with subsoil; and
- wasting soil by mixing it with construction waste or contaminated materials, which then have to be treated before reuse or even disposed of at landfill as a last resort.

The ease with which valuable and irreplaceable soils can be damaged means that careful planning is needed to protect this resource. Poor soil management will reflect directly on the landscape scheme, resulting in poor plant growth and failures, increased surface water run-off and ponding, or areas of water-logging. Creating and implementing a soil resource plan, setting out how to recover, store and reuse soils, is an important part of our work. This should include the storage areas for topsoil and subsoil, haul routes and how any stockpiles should be managed. Standards set by organisations such as the UK British Standards Institution (BSI) give useful guidance.

Trees

Most landscape architects will have had to patiently explain how vulnerable trees are to damage, that roots are generally shallow and broad rather than a single tap root and that the area under a tree is not the correct place to store materials. I have numerous photographs of poor practice on sites I have walked past, with barrels of chemicals or heavy machinery stored behind the tree protection fencing, ropes tied around a tree to act as leverage and overhanging branches damaged by

Fig 5.7 Despite their beauty trees are not always valued in landscape schemes – beech woodland, 2019, South Oxfordshire

machinery or vehicles. The potential impact of changing soil levels around a tree also needs to be discussed.

Getting the project team, particularly those working on site, to value trees that are to be retained can be a challenge. Our careful work to protect the important trees on a site, often with the support of an arboriculturist, can be undone in moments.

I felt this was such a central issue to our work I asked online if any other landscape architects had developed techniques to persuade their colleagues of the importance of trees retained on site and why they need protecting. The responses were insightful, as well as humorous. The level of interest suggests that this is an issue for many of us.

- *'Talk to them about their childhood and the value of trees to kids' imagination, enjoyment? Has to relate to personal experience /value.'* Rosie @Whicheloe
- *'I'd take them a copy of The Hidden Life of Trees. Just the notion of what they can feel and what happens that we don't & can't see might do the trick!'* Avra Ploumi-Archer @space_and_place
- *'Name them'* Wes West @wesayso
- *'Make a kindergarden/school adopt each tree. Kids will take care of them and also prevent from demolishing. Also as was mentioned you can name the tree* Maciej Kupczyk @KupczykMaciej
- *'1. they add financial value to a site, 2. increase peoples wellbeing by being there and even just in their memories. 3.They produce oxygen= healthier people, carbon capture... that's just the top three on my list!'* A Longley @Axlongley
- *'Say they're worth BREEAM points'* Tom Oulton @itsBIMupNorth
- *'Also mention it can help with Considerate Constructors Scheme scores (more BREEAM points) and awards for the project to be won. And partly education - induction slides, toolbox talks, also contract conditions which are regularly audited and the site inspected. A supply chain charter they also sign up to might help. Also consider signage in commonly used languages on site.'* Andrew Kinsey @AndrewDKinsey
- *'mention enforcement notices'* Louise Baugh @lpbaugh
- *'Put a life-sized picture of the Lorax on a wall in the office. But also, more seriously, have a chat about how trees take forever to grow to maturity.*

CITY OF TORONTO

The City of Toronto has a strong commitment to tree preservation, considered by some as ground-breaking. Their approach includes:
- punitive fines of up to $100,000 CAN (over £60,000) for damage to street trees
- describing tree works consent as 'a Permit to Injure or Destroy Trees'[5]
- a city-specific tree-protection policy for construction near trees, including the specification for a 2.4m high plywood tree protection hoarding
- a standard layout for tree protection signs within the city, including contact details for the department controlling permissions
- strict street tree protection by-law, including a ban on attaching any items, such as signs or decorative lights, to a street tree without approval
- a target of increasing the city's tree canopy from 27% to 40%[6]
- protection for all street trees and all private trees with a diameter of 30cm or more
- the requirement for a Tree Protection Guarantee deposit or letter of credit, 'which represents the appraised value of the tree to be protected, the removal costs and tree replacement costs' and is only released once the work has been approved by the City.[7]

It's always a shame to see a tree go'.
Lizbet_AmericanStone @MrsQuartzite

— *'Councils should give tree officers backing to shut down sites if trees are compromised. They have the power but frequently not the will or knowledge to enforce.'* Brian Hawtin @BJHawtin

The mention of third-party schemes such as BREEAM and the Considerate Constructors Scheme can be useful, as achieving accreditation might be an aim for the client.

Stressing the age and importance of the trees and emphasising the implications of the removal of protected trees, such as fines, can all help reduce bad practice. This should be backed up with physical protection, such as a fixed barrier, with weather-proof signage at regular intervals along the entire length of fence. As suggested by Andrew Kinsey, *the signs should be in all commonly used languages on site.*

The British Standard BS 837:2012 *Trees in relation to design, demolition and construction – Recommendations* is very clear about how any protected areas should be treated, stating that the area should be regarded as 'sacrosanct.'[8] It is a sadness that arguing for the retention and protection of trees is still part of our working practice.

Interestingly English planning law only asks that a tree provides 'amenity' for it to be eligible for legal protection.[9] The term isn't defined in law, but government guidance asks local authorities to assess visibility and individual, collective and wider impact including size and form; future potential as an amenity; rarity, cultural or historic value; contribution to, and

Fig 5.8 Out of season it can be hard to spot rare species – native bluebells in the Chilterns, 2019, Oxfordshire

relationship with, the landscape; and contribution to the character or appearance of a conservation area.

Other factors such as the importance to nature conservation or response to climate change may be considered, but they are not the central reason for protection. Providing habitat value, storing carbon and improving air quality are all valuable roles played by urban trees. However as landscape architects it is worth remembering that in some circumstances all that is asked of a tree is that it looks good.

Habitat

Depending on the nature of the site there may be a requirement to protect areas of important habitat, either within the site or in the proximity. This can be particularly difficult if the habitat is subtle, such as an area with a rare plant species that is only visible at certain times of year and hard to differentiate from the surrounding area.

Habitats can be vulnerable to many aspects of construction, such as dust, noise, light or vibration. A minute change in water quality or a surge of silt could destroy a complex and established aquatic ecosystem, or leached minerals from a footpath could affect the species in adjacent grassland, so it is important that measures are in place to prevent accidental damage.

Many of the projects I have worked on have included detailed habitat protection measures. One project, a scheme to create a new bypass channel to reduce flooding, included substantial work to protect the habitat of an important species of pea mussel. The mussel is only a few millimetres wide, so almost invisible to the naked eye. The ecology survey had identified the species and the scheme was carefully designed so the freshwater habitat wasn't affected, either during construction or in use. The project was commissioned by the UK Environment Agency, the organisation charged with the protection and enhancement of the environment, meaning that environmental protection was central to the project and that the whole project team understood the value of retaining important species.

However, on other projects with less environmentally aware clients habitat protection can be seen as an inconvenience. As with the protection of retained trees there is often legislation in place to protect important species, but as with retained trees if there is a need for enforcement by statutory authorities it is a sign that all other measures have failed.

Reputation

'It takes many good deeds to build a good reputation, and only one bad one to lose it.'
Benjamin Franklin

Once a project starts on site the risk of reputational damage, for the client and the supporting project team, is significantly increased. Setting standards and principles, such as those mentioned in earlier chapters including ethical sourcing of materials or assessment of labour practices in supply chains, can reduce some risks.

Reputation is a form of perception – it doesn't necessarily reflect the true nature of an organisation but it is how communities, suppliers, employees, regulators or customers perceive an organisation. For some organisations there is a gap between reputation and actual performance, either positive or negative. This is often described as the 'reputation-reality gap'.[10] Projects can sometimes reveal that an organisation is unable to meet the standards it has previously worked to convey, resulting in a change in perception that more accurately reflects reality. Eager marketing teams, keen to promote their company, can inadvertently raise expectations that can't be met. Other clients may be hampered by a poor reputation – perhaps due to inaccurate media coverage – and perceived as having little regard for the environment or for only being motivated by profit.

This gap can either be managed by improving the organisation's ability to meet expectations, or by reducing expectations by promising less. This might mean putting extra resources into staff training, or making sure timescales are realistic.

As the project moves from idea to reality – truly from idea to site – the role of the client changes. The consequences of design changes are far greater and the costs substantially increase. They are usually committed to completing a scheme in some form. The asymmetric relationship in favour of the client during the design stages now moves in favour of those delivering the work on site.

THE PROJECT TEAM

The composition of the project team may change at this stage. Specialist consultants whose assistance was required at the design stage may become less involved. The main change is the appointment of the contractor and any sub-contractors.

Design queries

Once a project moves to site design queries need to be resolved more quickly. Delays during the design stages might mean a delayed start, but with a live site the costs of delay are usually immediate. Despite this greater urgency the impact of each change needs to be assessed in full, and the wider consequences reviewed. A rushed decision could cause more problems later in the construction, or during the in-use stage of the project. In a crisis situation there may be a temptation to create a temporary solution and pass a problem forward, perhaps to a stage where the design team are no longer involved. In software development this is known as 'technical debt' – the impact of additional work caused by choosing an easy option in the short term rather than the best overall solution.[11] This is rarely a good approach unless the second step to correct the debt is agreed as part of the overall solution. You might use a visually appealing but poor wearing surface as temporary paving to make a site functional for an official opening but there must be an agreed permanent solution to follow when time allows.

Queries must be assessed against all the decisions made to date, such as design aims or items included to comply with planning conditions. Decisions and the rationale behind them need to be logged, and reviewed, so that future decisions are taken in line with that rationale. Undertaking careful negotiations with a planning authority to agree a planting plan, or a placement of a tree, only for that work to be undone with a snap decision on site, can have far-reaching implications and undermines our role in the project team.

Our work as landscape architects can be subtle, with parameters that are not always apparent, such as habitat conservation, visual impact in the wider landscape, protection of underground services or contaminated soil. This subtlety can make the placement of items or design choices seem arbitrary to the client or the rest of the project team, so it is important that the landscape architect assesses potential changes. We also need to ensure that we explain our reasons, ensuring that the work and experience required to make that decision is recognised.

The hierarchy of priorities that was agreed during the design stage, such as time having priority over

ALLIANCE CONTRACTING

Alliance contracting was created in the early 1990s as a response to the overly complex contractual relationships used for the construction of North Sea oil platforms, a relationship that often lead to project overruns and long-running legal disputes. The first notable alliance in the UK was used on the Andrew Oil Field for BP; this new form of contractual relationship led to such significant cost savings, reduced timescales and improved health and safety record that the concept was taken up by other sectors, including construction.[12]

With alliance contracting the traditional project structure of client–consultant–contractor is replaced with one contract between all parties.

A 2014 report produced by HM Treasury found that alliance contracting was best suited to complex projects, and depends on high levels of integration between all parties and 'committed and visible client and delivery team leadership to drive change and performance'.

The report describes an alliance as 'an arrangement where a collaborative and integrated team is brought together from across the extended supply chain. The team shares a set of common goals which meet client requirements and work under common incentives.'

In June 2018 the NEC4 Alliance Contract (ALC) was launched by the commercial arm of the UK Institution of Civil Engineers. This new style of NEC contract is designed for use on major projects or programmes of work, where long-term collaboration is central. It can also be used where a series of lower-value projects are combined to create a larger programme of work. The contract includes the requirement to 'act in a spirit of mutual trust and co-operation'.

Table 5.3 Differences in alliance contracting

ALLIANCE	TRADITIONAL
– One contract between commissioner and alliance of all parties	– Separate contracts between owner and each party
– Risk is shared between all parties	– Risk allocated to each party
– Success based on performance of all parties	– Performance judged individually
– Contract is written with an expectation of trust	– Contract is written with an expectation of disputes
– Contract defines outcomes – change and innovation are encouraged	– Contract is tightly specified to try to predict all outcomes – change not easily accommodated
– Based on trust and transparency	– Provision made for disputes
– Change and innovation in delivery are expected	– Difficult to accommodate change

Adapted from a summary by LH Alliances.[13]

cost, becomes more important during construction with the need for faster responses to queries. Delegation should also be agreed – what can be decided by the project team and what needs to be referred to the client or perhaps even to the client's board or senior managers?

Most clients would not want to be involved with the exact placing of paving items, as long as they meet the agreed style or concept, but they may want to know if a material type needs to change due to issues with supply. Agreeing this threshold is important. It can be financial, time or design based – the client may only want to consider changes where the cost increase is over an agreed level, the time implication is more than an agreed number of days or the departure from the original design is major. How any contingency is used and who manages the release of contingency is also an important issue that needs to be discussed.

Collaboration

The importance of a collaborative approach continues during construction. Collaboration means finding a consensus. It doesn't mean that everyone will be in perfect agreement but it also shouldn't mean that parties are grudgingly complying with a decision. The concept of 'serving the story', as mentioned in Chapter 4 , still stands.

Barriers to successful collaboration can range from basic issues such as incompatible file formats, badly managed document issue or poor layering standards in drawings, to harder to resolve issues such as incompatible behaviours or lack of competence in essential tasks.

Clarity of roles and responsibilities helps collaborative working, as does a problem-solving ethos based on solutions rather than blame. Problems will almost certainly occur, so ensuring that processes are in place to deal with these in a swift and professional manner is essential. The move from adversarial to collaborative ways of dealing with issues is reflected in some new forms of contract.

Honesty and project ethos

The need for 'mutual trust' is a requirement for all projects, whatever the size. Each member of the project team is reliant on some or all of the other members. You have to trust that work undertaken on site is to the standard specified, that the advice given is correct, that those representing you to others do so accurately and that the behaviour of all involved will not damage your reputation. Contractual arrangements provide a fall-back position if issues arise, but the day-to-day running of projects relies heavily on trust between all parties. Setting a culture of trust and transparency where individuals feel able to raise issues as they arise, however awkward, is a far better approach than sitting on problems hoping they will disappear.

Site management

The management of a construction site relies heavily on trust. We trust that others have completed work to the quality agreed. We trust that others have understood and acted on any health-and-safety risks, and will act responsibly. Although these human behavioural elements can't be removed, there are ways to use technology to support the process.

PROJECT MANAGEMENT AND COMMUNICATION TOOLS

Managing information within a project, managing communications and tracking decisions can be a complex part of managing a project. In 1994 the Latham Report ('Constructing the Team') described the construction sector as 'ineffective' and 'fragmented', recommending that there should be greater partnering and teamwork. In the intervening time technology has moved on and there are now tools available to improve team working.[14]

Despite these innovations the default method of communication within project teams is often email, a poor collaboration tool. Decisions can be hard to identify within a long email chain, and it isn't possible to hold real-time conversation. It also lacks accountability as not all systems log if emails have been delivered or opened.

There are several project management and communication tools that can be used in place of email, allowing documents to be linked to tasks or comments, and completed tasks to be marked as resolved. Originally described as Groupware, collaboration software can include features such as virtual meeting spaces, chat facilities and real-time sharing of documents. The main

benefit of these tools is that unlike email the messages become content that is visible to all, can be edited after posting and is searchable.

A 2012 report by McKinsey Global Institute (MGI) found that of the 4,200 companies surveyed 72% used internal social tools such as Slack, Yammer, Chatter and Microsoft Teams.[15] The results were impressive – those using the tools were 31% more likely to find co-workers with expertise relevant to meeting job goals. Time spent finding internal information or tracking down colleagues was reduced. MGI estimated that by fully integrating social technologies companies 'could have an opportunity to raise the productivity of interaction workers – high-skill knowledge workers, including managers and professionals – by 20 to 25 percent'.

Many of us probably use these tools already, such as shared electronic calendars or shared online documents, but in a disparate way that means there are numerous locations to monitor. The best tools combine all communication and document management in one place, allowing users to receive notifications in a way that suits them, but ensuring that there is a central point for reference.

However, they only work if all parties use them – they need to be the only source of information or users default back to less open systems such as email.

Non-sector-specific examples include:
- Airtable – online spreadsheet database hybrid which can store images and other attachments, allows multiple users to work simultaneously on the same file, or base, and integrates with email, calendars and social media.
- Basecamp – web-based project collaboration and management tool, which includes to do lists, calendars, file sharing and forum-like messaging.
- Google Apps – web-based real-time document collaboration.
- Microsoft Office – includes tools for real-time document collaboration.
- Slack – cloud-based collaboration tool using a chat format which integrates with many other tools, such as Dropbox, Adobe Creative Cloud and Google Drive.
- Trello – web-based project management application that organises projects into boards, providing a visual representation.
- Yammer – team collaboration software and private social network.

PACMAN

When Bristol-based civil engineering and scientific consultancy Edenvale Young Associates failed to find project management and time-recording software that met their needs, they commissioned their own bespoke management software. Their first version, developed as part of their ISO 9001 Quality Management System (QMS) accreditation, was created in 2014.[16]

This then evolved into PACMAN, a complex database that is viewed in a web browser. The system integrates:
– client data, including contact details, project agreements, invoicing, expenses, change control and customer feedback
– assignment of project managers, sales management including the cost of preparing proposals and projections of future workloads
– timesheet entry, staff costs, staff utilisation, overtime, annual leave and expenses
– project-specific milestones such as tendering, start-up, issue and completion dates
– supplier and sub-contractor information, including invoicing and payments.

Combining all this information in one place allows Edenvale Young to create a series of dashboards allowing them to:
– examine the daily status of live and completed projects
– track invoicing and payments
– manage workload and staff utilisation
– identify projects that need careful monitoring.

The system allows rapid auditing in accordance with ISO 9001. An interesting feature of PACMAN is the 'Improvements' section – a place where staff can post ideas or identify problems as they arise.

We can use technology, such as smartphones, to record progress on site and to create a log of work completed. An inexpensive webcam can be used to monitor specific site locations, such as the access points, allowing the site to be viewed in real time or for the saved imagery to be used as evidence of deliveries or site visits. Photos can be saved to shared online albums or to project management software such as Trello or Basecamp. Video calls using apps such as Zoom or Skype allow site staff to show rather than explain issues, allowing technical problems to be solved by off-site colleagues or suppliers. Privacy and security issues need to be complied with, but it is possible to use the technology that many of us use in our personal lives to make our working lives more efficient.

More complex technology can help monitor progress, such as augmented reality headsets or cloud-point scanning.

Technology is a useful adjunct to support our work on site, but in no way replaces time spent in person, talking to site staff and seeing the work as it is completed. This work needs to be reflected in our fee quote, which can be a challenge to argue but the quality of work undertaken reflects on our reputation – not having the time in the fee budget to check work and to be involved in agreeing design changes can mean decisions are made without our input but the landscape scheme is still attributed to us.

If the work is complex or unusual, time needs to be allocated to explain the principles behind the design, and why elements have been laid out in the way you have specified. The person working on site is unlikely to have been involved in the design process, and a 2D drawing does little to explain the rationale behind a design. Technology is a huge help but nothing replaces building a good relationship with those turning your project from design into reality.

Sourcing materials

The supply and sourcing of materials can be a risk to progress during construction. Landscape architects have the added complication that some of the materials we source are living, and take time to grow, so can't be manufactured on demand like other components. There is a finite supply each year of trees of a certain size in each species, and issues such as poor weather can reduce availability. Release of plant stock can also be affected by the weather – nurseries only release their bare-root plant stock once the plants have experienced weather cold enough to trigger dormancy.

Reserving plant stock in advance is a sound move, especially for rare or larger specimens. Visiting a nursery during the summer, when the plants or trees are in full leaf, makes it easier to check health and to select trees of a similar shape or form for planting in groups or avenues. Once the individual plants or trees are selected at the nursery, tamper-proof tags are usually fixed to the tree and the numbers recorded, so the landscape architect can be certain that the trees they selected are the ones that arrive on site.

It can be hard to persuade a client of the value of a nursery visit, especially if it involves the time and cost of travelling outside your own country. However,

unlike mass-produced building components, seeing one sample of a tree won't ensure that those supplied will be of the same form. We need to help our clients understand the natural variation of size and form that can occur within one tree specification. The UK British Standard *BS 3936-1 Nursery stock-specification for trees and shrubs* allows 50cm variation in the height of a Standard tree, which can make a noticeable difference to the size of the canopy, and mean that trees of the same specification may not look balanced in a group.[17] This might not be an issue in an informal park or a habitat site, but would look out of balance in an avenue or as an entrance feature.

As the project moves from concept to reality our role can change. If we are a long way down a sub-contractual relationship we may not even get to site – a risky situation as design changes can be made with no understanding of the rationale behind our decision. A distant relationship, with no direct access to the client or site staff, makes it more difficult for us to deal with issues. If we are leading the project or part of a project team where each practitioner has equal status we may spend a substantial amount of time on site, checking details, building relationships and dealing with issues as they arise.

WIDER SOCIETY

Once construction begins the physical impact on the places and communities we work in becomes tangible. The site is cleared, the site compound is set up and the lorries start to arrive. However carefully the consultation process or publicity has been managed

MIXED REALITY ON SITE

Mixed reality, a development of augmented reality, is a technology that adds computer generated content over the reality around you, with the virtual objects anchored in the real world. Mixed reality is an advanced form of augmented reality, a technique used by the computer games sector in apps such as Pokémon Go. Unlike virtual reality, where the user is immersed in a completely virtual environment, mixed reality allows users to see the site as they move around, and to call up information linked to the real world objects in front of them.

With virtual 3D models of the project being created as part of BIM, mixed reality is now being used in the construction industry. Wearing devices such as Microsoft HoloLens, either as smartglasses or integrated into a hard hat, users can see the 1:1 scale virtual model over the existing site.[18] Data linked to the virtual model, such as installation instructions, can be accessed using eye movements, gestures and voice commands.

This hands-free system is similar to the heads-up display system used by pilots to allow them to view relevant information whilst looking up, rather than looking down to check instruments. The smartglasses don't have wires or require a phone or PC connection, allowing site staff to access information hands free as they undertake their work. The system can be used to check defects and to create a snagging list linked to the virtual model.

Landscape architecture often involves the careful integration of new elements into an existing setting. Mixed reality could allow us to look at the visual impact of a project on site and in real time. There are already tablet-based tools that allow users to test ideas from site, creating simple massing models or overlaying basic ground modelling plans, but these are quite limited in scope and require the user to hold a tablet at eye level.

Such systems are still expensive – at the time of writing the Trimble Xr10 with HoloLens 2 had a recommended price of US $4,750 – but like all new innovations it is likely that that the size and the price will reduce over time. I look forward to the development of more discreet options, as the idea of a hands-free device that allows you to take photographs, view a site plan or test viewpoints, usable with gloves on cold site visits, is an appealing prospect.

there will still be those for whom the works are a shock, and their concerns need to be dealt with tactfully. It isn't us who is woken by a reversing lorry arriving on site as we are trying to catch up on sleep after a night shift, or irritated by the noise and vibration from machinery as we try to work from home. They may be looking forward to the completed scheme, and the positive benefits for their community, but in the short term the disruption can be hard to live with.

The public's role in construction

Many of the projects we undertake are at least in part for public benefit, either explicitly as an area for public enjoyment or more implicitly by providing environmental benefits. The public may not be our direct client but there can be very few projects that have no impact, either positive or negative, on the wider world.

The American Society of Landscape Architects (ASLA) Code of Professional Ethics states that

'Members should endeavor to protect the interests of their clients and the public through competent performance of their work',[19] making the point that our work has a wider impact that we must acknowledge. I take this to mean that we should consider the public at all stages of our work, from inception to demolition, and in particular during construction when many of the potential impacts can arise in a short period of time.

Managing disruption

Our work often has an adverse impact on those who live and work near the sites as they are constructed. We may love our work and enjoy seeing our designs take shape but the activities needed to create our schemes can be a real intrusion on people's lives. This potential for disruption needs to be considered at the design stage, as some more intrusive practices can be designed out.

A paving scheme with minimal need to cut blocks on site will not only be better for the health of the site staff, it will also reduce noise and dust. Soil management can often be planned so that the soil on site is retained, with minimal need for soils to be removed or introduced. Reviewing site processes and looking at whether more disruptive aspects can be avoided, or measures taken on site to minimise impact, can all reduce disruption on site.

Updating the public

Making sure that residents and businesses have the information they need is an important part of many construction projects, allowing people the opportunity to plan around the most disruptive spells.

Open days – For projects that impact on people's home and work lives holding public open days be a useful way to address concerns. There may be substantial local interest in a scheme so allowing a behind-the-scenes visit during construction can be a useful opportunity to talk to local residents in an informal setting. Residents may bring up issues that they wouldn't mention in a more formal setting, such as site staff behaviour or noise issues. As in the consultation process this shouldn't be a tokenistic event – it should be a genuine gesture. There may be issues that are negligible to us but of concern to local residents that we can easily resolve, or there may be factors that there is no way to mitigate. If we don't have a discussion we will never know.

Open Doors Week, held in England, Scotland and Wales each March, gives the opportunity to see behind the scenes on live construction sites including road schemes, hospitals and science parks.

On-site information – Information about the work can be communicated in a number of ways, depending on the nature and location of the project. For a minor scheme a poster on site, outlining the work planned and listing contact details, might be adequate. For complex or controversial projects there may be a greater need to manage publicity, responding to calls, emails and social media enquiries as well as media coverage. At the very least all sites should have details of the client, the duration of work, a description of the work and emergency out-of-hours contact details. Other details that could be included are contact details for the project team, evidence of any permissions granted and contacts for these, upcoming events, links to online resources, details of funding organisations and the background to the project.

WHAT3WORDS

One of the major frustrations for those living and working close to construction projects is the number of vehicles, blocking traffic as they deliver to site or turn in a narrow street.

For many locations the delivery address might be a postcode, but the location can be ambiguous. Even with a street number the entrance to a site can be difficult to find. In some countries there may be no addressing system in place.

One solution to this problem is the addressing system what3words.[20] The system divides the surface of the world into 3m x 3m squares and assigns each square a unique three-word address, that is easy to remember and works well with voice input on devices such as mobile phones. For example the southern pedestrian entrance to Paddington Station in London is ///peanut.plenty.ideas and the summit of Snowdon in Wales is ///museum.outsiders.ballooned

Uses in our sector could include mapping the location of assets such as trees or street furniture, creating an address for a site entrance or locating features in open countryside. The system is integrated into the open-source GIS software QGIS, discussed in Chapter 3.

What3Words was created by music-industry professional Chris Sheldrick in response to the fact that the bands, equipment and supplies he booked for live music events frequently got lost. He worked with two friends to develop the system, which is now available as a mobile app, a photo app and an online map tool. The system is used by organisations including the United Nations who use it as part of their disaster reporting app.[21]

Fig 5.9 what3words address for site in Wales

It is worth reviewing how the public might perceive the work, especially as images and information can spread quickly via social media. Any techniques that could be mis-judged may need to be explained. In the UK a recent photo of netted trees around a construction site in Guildford was shared over 2,500 times, liked over 4,000 times and received over 1,800 comments in just 36 hours, and led to articles in the national press.[22] The poor image of construction means that the public often assume the reasons are negative rather than positive, so providing an explanation on site is one way to counter this risk.

Emergencies – At the very least all sites should have a 24-hour emergency contact number that is always answered. The number doesn't need to be a specific telephone line – a local or freephone number can be bought for the duration of the project then diverted to either a duty mobile or an out-sourced switchboard who can then forward calls to the relevant team member.

For all but the smallest schemes project teams should set up a process for dealing with public and media enquiries, agreeing who is the lead contact. An informal call from a journalist on a Sunday afternoon,

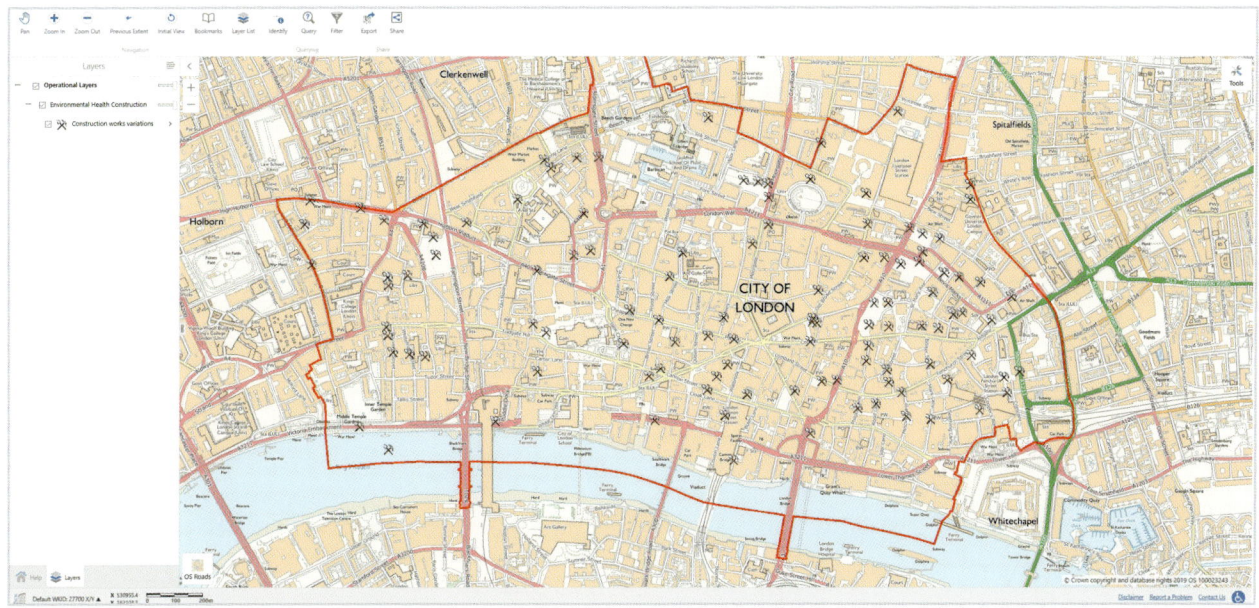

Fig 5.10 The City of London interactive map allows residents to check for approved out-of-hours construction work, 2019[23]

if not carefully managed, can escalate into a story that may not show the project or the client in a good light.

Journalists, bloggers and social media communities can be useful supporters for projects but bad press coverage can be hard to counter, and can rebound on any team member. Having an agreed and experienced point of contact who everyone knows to direct enquiries to, and has their number saved in their contacts list, is one way of ensuring a consistent approach. A Frequently Asked Questions (FAQ) page, either on the project website or shared with the project team, can be a useful way to manage queries from the public.

When we begin work on site we need to remember that we are working in areas that people live and work, and that even projects with strong local support can try the patience of those who are adversely affected. Work during the design stage can mitigate some of the potential disruption but we also need to manage the processes that can't be avoided and ensure that the community is kept updated.

Moving the project from plan to site is an exciting step. It is the point where hidden issues come to light, and the design is tested. The risks become greater, as money is spent at a far greater rate. Once the work is complete as-built drawings are issued, defects rectified and maintenance programmes agreed.

CONSIDERATE CONSTRUCTORS SCHEME

The UK non-profit Considerate Constructors Scheme (CCS) was founded in 1997 in recognition of the need to improve the image of the construction industry. The Scheme was created by the Construction Industry Council, in response to the findings of the Latham Report 'Constructing the Team', which described the construction industry as 'ineffective' and 'incapable of delivering for its customers'.

The Scheme recognises that 'If all construction sites and companies presented an image of competent management, efficiency, awareness of environmental issues and above all neighbourliness, then they would become a positive advertisement, not just for themselves but for the industry as a whole.'

Construction sites, companies and suppliers can register and they must meet the terms of the Code of Considerate Practice. Sites are visited by Site Monitors and the Scheme includes a process for managing complaints about sites.[24]

The Code is made up of five parts:
- care about appearance
- respect the community
- protect the environment
- secure everyone's safety
- value their workforce.

The section on respecting the community states that 'Constructors should give utmost consideration to their impact on neighbours and the public':

- Informing, respecting and showing courtesy to those affected by the work.
- Minimising the impact of deliveries, parking and work on the public highway.
- Contributing to and supporting the local community and economy.
- Working to create a positive and enduring impression, and promoting the Code.'

As part of the assessment process CCS Monitors ask 'Are all those affected by the work identified, notified and kept informed and shown courtesy and respect?' which includes the provision of 24-hour contact information for the public, and a standard signboard is available to show the number.

Their Best Practice Hub is a useful resource, including ideas for managing

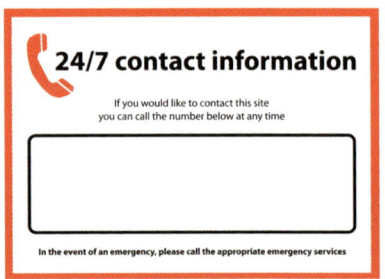

Fig 5.11 Considerate Constructors Scheme poster for 24-hour contact

site entrances and site signage. The Community section provides useful case studies, with ideas including free professional car and window cleaning for neighbours affected by site dust, training a dementia champion to ensure a dementia-friendly site and creating a system for visual community for students with autism.

Fig. 5.12 Comprehensive site signage showing contact details and updates for neighbours; Morgan Sindall Construction & Infrastructure Ltd for Monmouthshire County Council 2015, Raglan Primary School, Monmouthshire

Case Study 5.1
NEW LUDGATE
CITY OF LONDON

Title
New Ludgate Piazetta

Client
Land Securities Group (Landsec)

Location
16 Limeburner Lane, Ludgate Hill/Old Bailey,
City of London, UK

Year completed
2015

Project Value
£2 million

Type of Scheme
Public realm and roof garden

Landscape Architect
Gustafson Porter + Bowman

Owner
Land Securities Group (Landsec)

Contractor
Main contractor - Skanska

Suppliers
Plant supplier - Coblands, Deepdale, Lindum, Robin Tacchi Plants, Van de Berk,

Paving - Marshalls, Gormleys

Fig 5.1.0 (Page 173) Bespoke stone benches; Gustafson Porter + Bowman for Land Securities, 2019, New Ludgate, London

Fig 5.1.1 (Page 174) Belle Sauvage Passage looking towards Limeburner Lane, showing close-up of paving detail; Gustafson Porter + Bowman for Land Securities, 2019, New Ludgate, London

Project Team
Structural consultant - Waterman Group

Masterplanning - Fletcher Priest Architects

Planning consultant - DP9

Architectural - Fletcher Priest Architects, Sauerbruch Hutton Architects

Quantity surveyor - Gleeds

Mechanical and electrical - Waterman Group

Lighting consultant - Speirs & Major Associates Ltd

Fire consultant - Arup

Access consultant - BuroHappold

Awards
2017 Winner, The Chicago Athenaeum International Architecture Awards

2017 Shortlist, RIBA London Regional Award

2017 Shortlist, Architects' Journal Architecture Award (landscape category)

2016 Winner, City of London Building of the Year Award

2016 Winner, LEAF Awards Developer and Development Project of the Year

2016 Winner, RICS Commercial Office Building of the Year

2016 Highly commended, Landscape Institute, Best smallscale development

2016 Shortlist, NLA Office Building of the Year

2015 Shortlist, BCO Commercial Workplace Award

2015 Shortlist, FX Awards Public Space Scheme

Named after the westernmost gate of the Roman London Wall, New Ludgate is close to the Old Bailey, the Central Criminal Court of England and Wales, and St Paul's Cathedral in the City of London. The history of the site reflects many of the major events in London's history – the area is shown in the scorch maps of the Great Fire of London in 1666 and in the maps of bomb locations in the Second World War.

After the dense collection of shops, yards and public houses was destroyed by bombs the site was rebuilt, and by 2011 was in need of redevelopment. A 1980s office block provided limited public amenity and breached Primrose Hill and St Paul's Cathedral viewing corridor height restrictions so Land Securities Group (Landsec) approached the City of London, the planning authority for the borough, to discuss the redevelopment of the block.

London based landscape architecture practice Gustafson Porter + Bowman were appointed at an early stage in the project, appointed directly by the client and working on an equal footing with other members of the project team. The landscape team were involved in early design workshops ensuring that the landscape scheme was included at the masterplan stage.

The masterplan for the site adopted the concept of the narrow lanes typical to the City of London, placing two nine-storey buildings either side of a new public thoroughfare, named Belle Sauvage after the 15th-century coaching inn which stood on the site and which was demolished to make way for a Victorian railway viaduct. This accommodated the change in level across the site as well as providing improved pedestrian access. The porters and messengers using the passage

Fig 5.1.2 St Martin, Ludgate, Ludgate Hill, City of London: the steeple seen from the south-west with St Paul's Cathedral in the background, 1896

as a shortcut provide a reminder that the area has been a busy commercial district since Roman times.

The site is in a high-density area, with offices, shops and cafés, but there are limited opportunities to sit outside. The site is privately owned and managed but the client was keen to blur the boundaries between public and private space.

In many locations in the City of London a slot drain or change in paving provides a clear demarcation between ownership. Wanting to avoid this hard visual boundary the design team created a paving pattern that created a transition between the two spaces. Gustafson Porter + Bowman used hand-drawn sketches and 3D modelling to develop a set of 9 repeatable

Fig 5.1.3 Belle Sauvage Passage looking towards Limeburner Lane, showing paving detail; Gustafson Porter + Bowman for Land Securities 2019. New Ludgate, London

The landscape team were heavily involved with work on site, spending time with site staff to ensure that they understood the concept behind the design and that they knew who to come to with queries, however minor. Knowledge of the NEC contract was invaluable, as it allowed the landscape architects to be closely involved in decision making. They were also involved in sub-contractor selection and pre-approving items with the main contractor.

The default reaction in construction when an issue arises is that someone has done something wrong, and that blame needs to be placed, but the team were keen to foster a more collaborative ethos. Their approach was to not react hastily, but to stand still and wait for the full situation to come to light, allowing others the time to explain their role. However carefully a project is detailed there is still a need for numerous decisions on site, especially with landscape projects with unknowns

paving shapes. The stone paving, cut using the 3D data, appears as a random pattern with an increase in complexity towards the centre of the site, reflecting the transition from public to private space.

At street level the stone benches in the small piazzetta provide a quiet place to sit under a tulip tree (*Liriodendron tulipifera*) away from the main street. The shape of the bespoke benches is extruded from the paving pattern, creating seats at various heights that also act as a subtly integrated barrier to vehicle-borne attacks.

At a higher level the project includes a roof garden, utilising the step in the building outline required to maintain sight lines to St Paul's Cathedral, and two areas of green roof.

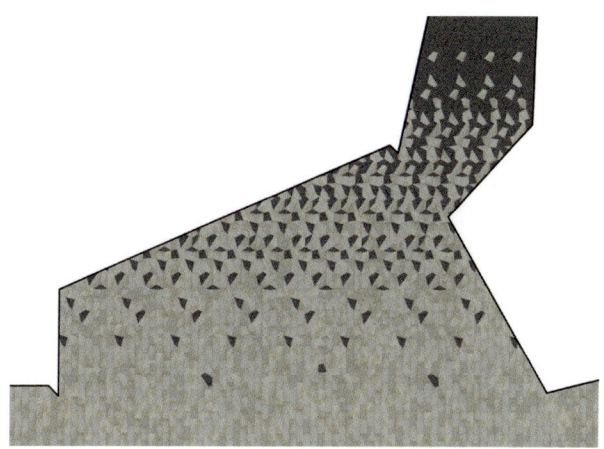

Fig 5.1.4 Plan of paving detail; Gustafson Porter + Bowman for Land Securities

such as weather or the variation inherent in natural materials, so a regular presence on site is essential.

The boundary between public and private space brought practical issues on site, as the paving outside the client's ownership had to tie into the design but was implemented by a City of London contractor. Gustafson Porter + Bowman worked with both contractors to ensure the blur between the two areas of ownership was achieved.

The client wanted the project to be site specific – the landscape team's response to this brief was a complex and unique space that demonstrates the value of involving landscape architects early in the design process and allowing them equal status with the rest of the design team.

Fig 5.1.5 (Page 178) Bespoke stone benches echoing the paving pattern; Gustafson Porter + Bowman for Land Securities 2019, New Ludgate, London

Fig 5.1.6 View of piazzetta tree and benches, looking outwards to Old Bailey, Gustafson Porter + Bowman for Land Securities 2016, New Ludgate, London

Fig 5.1.7 Entrance to Belle Sauvage Passage from Limeburner Lane, showing the absence of a boundary line, Gustafson Porter + Bowman for Land Securities 2019, New Ludgate, London

Fig 5.1.8 (Pages 180 and 181) Roof garden; Gustafson Porter + Bowman for Land Securities, 2019, New Ludgate, London

Fig 5.1.9 (Page 181) View of piazzetta; Gustafson Porter + Bowman for Land Securities, 2019, New Ludgate, London

Fig 5.1.10 (Page 181) View from roof garden to St Paul's Cathedral, Gustafson Porter + Bowman for Land Securities, 2019, New Ludgate, London

Chapter Six

MAINTAIN AND EVALUATE

INTRODUCTION

With the work on site finally complete we reach the stage where we fulfil the client's brief and finally produce a site that can be used. When we are busy dealing with the day-to-day complexities of running a project we can sometimes forget that this is what we have been working towards. For us it might be the end of our involvement but this is the stage the client commissioned us to achieve. As our work comes to an end the site comes into use, a stage all the proceeding steps have been leading up to.

The use of living materials in our projects means that maintenance is essential as trees and plants establish. Even low maintenance does not mean no maintenance – like any living things plants need care to survive, especially as they establish. The most carefully managed planting scheme is likely to need plants replaced as they fail. We need to ensure that clients understand the need for, and commit to, a suitable maintenance regime.

This is also the stage for review, as reflected in the RIBA Plan of Work. The inclusion of feedback and project review was an addition to the 2013 Plan of Work, emphasising the importance of evaluating our work. The updated plan shows the sequence of stages as a circle, representing the idea that experience from the end of one project should feed into the start of the next. For landscape architecture this cycle is apt

– our work is always an ongoing cycle of review as a project evolves.

THE LANDSCAPE ARCHITECT

As the work on site is complete our involvement in the project may reduce. If we have done our job well we might have designed ourselves out of the process, leaving our client with a sustainable maintenance plan that allows them to manage the site day to day. If we are exceptionally lucky our client may see this stage as the start of a longer process and involve us in the long-term care of the site, helping them refine the design as new opportunities or issues arise. It is a luxury to work with a client who sees landscape architecture as an ongoing process that is constantly

STAGE 7 – IN USE

- Manage completion of tasks in handover strategy
- Manage updating of project information
- Carry out site inspection
- Carry out site survey
- Project team meeting
- Undertake tasks listed in handover strategy
- Design team meeting
- Carry out design/technical review
- Review updated project information
- Declare carbon performance.
- Exchange updated 'as constructed' information
- Manage maintenance operations

Table 6.1 The elements of this workstage[1]

Fig 6.0 (Page 182 and 183) Sheep grazing in front of Skiddaw – the landscape of the Lake District fells is managed by grazing, 2011, Skiddaw, Lake District National Park

Fig 6.1 Example of poor maintenance – park bench with moss, 2019

Fig 6.2 Restoration work at Parc Cefn Onn, funded by the National Lottery Heritage Fund; expert adviser Claire Thirlwall for National Lottery Heritage Fund Cymru 2019, Cardiff

reviewed and refined, rather than a means to an end.

It is also the point when we can review all the workstages and evaluate not only the project but the performance of the entire project team, including ourselves.

If we don't evaluate our work we have little idea whether our project was a success. Our completed work is the visible result of all our work, and is how we will be judged by colleagues, clients and the wider public. Much of our working practice is hidden, such as how we conduct ourselves in meetings or the quality of our drawings, but the final result of all our work is on show for all to see. We can also review whether we have met the standards we set for ourselves.

Maintain

There is little point in writing a project brief, designing a scheme, consulting stakeholders, gaining statutory consents, spending often large amounts of money and constructing a project if it isn't going to be maintained. This may seem an obvious point but most of us know neglected sites with evidence of a good landscape scheme that have suffered from a lack of maintenance. These might even be sites that we have designed and been unable to prevent falling into decline.

This should not be the first point where the issue of maintenance is discussed – the resources available to maintain the site should have been agreed at an early stage, as the choice of materials has an impact on the level of maintenance required. Decisions such as the type of grass seed mix used or the edging for pathways are dependent on the expected level of maintenance.

Few clients can predict with absolute certainty the budget they will have available to manage a site, and public-sector organisations are always vulnerable to budget cuts, but an allowance for maintenance should be part of the project budget. In the UK planning consent often requires that a site is maintained for at least 5 years and some funders such as the National Lottery Heritage Fund ask for financial resources to be provided for 10 years after completion, commitments that the client needs to accommodate.[2]

A process needs to be in place for unforeseen or irregular costs, such as clearing graffiti or vandalism. The process also needs to cover long-term maintenance tasks, such as de-silting ponds, repainting railings or thinning of tree planting areas. Given the potential longevity of the schemes we design there may be a need to look at a timescale of tens or even hundreds of years. Not all clients will be thinking so long term, but established organisations such as charities or private landowners whose family have owned a site for generations may consider maintenance within a much longer time frame. Historic landscapes need to be managed at this timescale to retain their landscape character, for example by looking at succession planting of significant trees, as even the largest tree from a nursery is tiny in comparison to a mature parkland tree.

Maintenance – why is it needed?

Good maintenance reduces the risk of plant failures, but plants are living material and susceptible to pests and disease, not all of which are treatable. Techniques for reducing the impact of pests and

Fig 6.3 Tree damaged by poor maintenance – tree guard still in place 20 years after planting, 2019, UK

disease are discussed later in this chapter, but managing a landscape scheme is a commitment for the entire life of the project. Even the most naturalistic sites such as new habitats need to be managed, particularly if they are in an area where non-native invasive species are present.

The closer a site is managed to the type of habitat that would ultimately colonise that site, known as the climax community, the less input is needed for maintenance.[3] For example, keeping a site that would naturally become deciduous woodland as bare earth would take a high level of resources – think of commercial farming and the work required to keep the soil weed free, or the time needed to manage beds of annual plants. The site can be adjusted to reduce the level of input – the Merton Borders Case Study in Chapter 4 are seeded into a layer of low-nutrient sand, conditions more suited to the prairie planting that the beds emulate. Planting styles that mimic an ecologically balanced plant community may require less input, but they will never be self-maintaining and will always require some care.

For many of our sites the role of maintenance is to prevent the steps that would lead to a site becoming a climax community – the first appearance of pioneer species, usually small species that are tolerant of extreme environments and short lived. They often help to form soils and modify the environment in ways that supports the next level of habitat succession. In our maintenance plans they are described as weeds but they demonstrate the beginning of a slow progression towards the site reaching an ecological balance.

THE LONG WALK, WINDSOR

A useful example of succession planting and the slow dynamic of landscape architecture is The Long Walk. The broad avenue that leads up to Windsor Castle was originally planted with in the 1680s with a double row of elms. The majority were felled as they reached the end of their lives and replaced from the 1850s onwards with a mix of horse chestnut and London plane, with the remaining elm removed in the 1940s due to Dutch elm disease. This iconic landscape, with careful landscape management, has survived over 300 years.

Unless the site is remote wilderness with no human access, no imbalance of species that requires population management – such as deer – and no threats from pest and diseases it is likely to need some form of management. This may be checking the health of trees to protect the public from injury through falling trees or branches, or managing animal populations with physical measures such as fencing, but unless the site is in an area where the ecology is fully balanced some form of maintenance will be needed.

Management and maintenance plans

A review of landscape management and maintenance plans submitted as part of UK planning applications suggests that they are a usually a fixed format, such as a long document produced via a word processor, perhaps with tables to summarise the tasks in different time frames, issued as a PDF.

This works well when the role of the management and maintenance plan is evidence for a planning application, as it is simple to review and in a commonly used file format.

However, with the developments in software for landscape architects it isn't a huge step for our digital drafting software to generate maintenance plans in parallel with plant and material schedules. With the advent of BIM, items within drawings, such as plants or street furniture, now have data attached to them. The complexity of landscape maintenance, with different maintenance options even for the same species of plant – maintenance of a hornbeam hedge is different from a specimen tree – means that creating a standardised list of maintenance terms is difficult. Despite the complexity the concept is possible, and will be a welcome development.

As well as automatically creating maintenance plans as we design a scheme, the estimated maintenance costs could be calculated to give the client an indication of whole-life costs, in turn helping persuade clients that higher upfront costs could be offset by long-term maintenance savings, such as higher-quality paving materials or mowing strips to grassed areas.

Evaluate

Once a project moves into the maintenance stage and begins to establish we can review the choices made, such as plant types and hard landscape materials. Each project we work on should teach us something that we take into our next piece of work. Whether we formally hold a debrief to discuss the project as a team, undertake in-depth research or just take a few minutes to reflect in between the urgent items on our to do list, evaluation is an important step.

If we don't evaluate our work we don't know if we achieved the original aims of the projects, aims that can get lost in the day-to-day work of getting a project to site. Landscape architects are used to evaluation at the early stages of a project, looking at visual impact, landscape character or habitat impact to help make decisions about the design, but that doesn't then always continue through the whole life of the project. The purpose of evaluation is to prove and improve – to prove that the project fulfilled its original aims and to justify the client's spend, and to improve future work.

Having an approach that constantly looks for improvement, and is open to new ideas, can help find time savings in workflows or improve the way we work with clients and colleagues. Assuming that what you do now is the best or only approach, and never taking the time for review, can mean that you retain outdated working practices. It doesn't mean giving up on successful techniques just to adopt the latest idea, but taking a critical look at your workflow can at the very least reduce routine tasks and free up more time for the more enjoyable aspects of our work.

Evaluation of a project may reveal processes that could be improved, such as the use of technology to share documents, or practices that were successful such as site open days or stakeholder consultations. One simple point to evaluate for commercial projects is whether the project made a profit, and if not which elements caused the loss.

My experience is that what often takes a project over the time budget is the additional day-to-day project administration, such as telephone calls, meetings and email exchanges, when a project is delayed. Each task in itself is small, so doesn't warrant a request for additional fees, but if a project extends a few months over the original timescale the time taken just to be involved in the project can be substantial. Including predicted timescales in fee quotes and a rate for any overrun can be one way to address this issue, but this approach can be hard to implement when the overrun is in small, unexpected increments. Careful time recording, including the type of task undertaken, means that unprofitable projects can be reviewed and the use of time analysed, in turn helping to improve the accuracy of future fee quotes.

Evaluating your own performance is also important. Poor communication, rather than lack of skills, is often the weak point in projects, with a reluctance to raise issues or an assumption that the landscape architect or other team member doesn't need to know certain details. Setting up systems to improve communications can help, but an ethos of trust within a project team and an understanding that all members of the project team are equal is even more important. Project teams can develop a hierarchy that was never intended by the client, with some members gaining a disproportionate influence. Evaluating our role in the team, how our advice was regarded and whether we had an equal voice in decision making may help us decide if we wish to work again with that client or with members of a project team.

Feedback – ask the client

Online feedback is now commonplace in many sectors. After booking a holiday, making an online purchase, arranging car repairs or even attending a doctor's appointment you might be asked to rate the service you have received. The rise of sites such as TripAdvisor and TrustPilot, along with the customer reviews on sales sites such as Amazon, has changed the way we plan our purchases – Amazon, a site that has allowed customers to leave reviews for their purchases since 1995, is now one of the most important reference points for consumers, even if they go on to buy the product elsewhere.[4] Fake reviews can skew open, unverified review sites, but they are still a useful starting point for many consumers.

For potential clients looking to work with a construction professional there doesn't appear to be an equivalent review process. Some practices have Facebook reviews and there are a number of landscape practices listed on the anonymous employee review site Glassdoor, but third-party reviews are rare. That may not be a bad thing, as open feedback is easily manipulated and is best suited to high-volume items, where the fake or malicious reviews are offset by genuine statements. Review sites for verified tradespeople exist, but they don't extend to construction professions. In the UK the Landscape Institute has a list of registered practices and the option to check the membership status of qualified landscape architects but there is no option for rating practices.

It will be interesting to see whether changes in working practice, perhaps along the lines of the gig economy mentioned in Chapter 2, will begin to impact on our work – the website freelancer.com has a ratings system, with a feedback section showing the project value, a review and a star rating for different aspects of performance.

The work we undertake is more complex than a single transaction, but it is worth considering how we could gain feedback. How do our clients let us know their views on the project? Do we ever ask the outright question, or do we assume that the absence of complaints is enough? Do we ever ask the client what worked, and what didn't work?

With a small organisation it can be awkward to ask for feedback from clients as it is often a close relationship, and we might fear opening up issues that have been resolved. However uncomfortable a process I believe we do need to find a way to ask for feedback. Not knowing about a problem doesn't mean that the problem isn't there. We might use technology such as an online survey to make the process more detached, or employ a third party to ask on our behalf. Or we could just ask.

Some multi-disciplinary practices undertake regular client consultations, with a score that is monitored and action taken if the score falls below a target level, allowing the project to be refined as it progresses. This may not be possible for smaller businesses but finding a way to allow clients to give genuine and honest feedback can help improve the relationship and future work with them.

Revisiting your work after a number of years can be enlightening – seeing how woodland areas have grown and which species, if any, have dominated, how people move through an established site and whether their

KAIZEN

Kaizen is the Japanese word for improvement.[5] In business the term refers to the process of constant, continuous improvement, with employees at all levels encouraged to make suggestions for improvement at any scale, with even the smallest improvements considered on the basis that in combination they can lead to major benefits.

The concept extends to customer feedback – Japanese retail company Muji has a Kaizen section of its UK website where it asks customers to offer suggestions, report problems and recommend additions.

Having a stated ethos of continuous improvement makes it clear to clients that you are open to suggestions and that you are constantly refining your working methods.

behaviour has modified the design, such as desire lines or materials that have not aged as expected.

Testimonials from clients and colleagues are a good way to promote our work, as statements by others are given more value than statements we make about ourselves, a concept known as social proof.[6] It provides verification, even if the reader doesn't know the reviewer. Again, this highlights the value of asking for feedback but does require the confidence to ask the question.

Continuing professional development

A commitment to continuous improvement and learning is central to many codes of conduct for landscape architects. All corporate members of the UK Landscape Institute must complete a minimum of 25 hours per year of continuing professional development (CPD) as a requirement of membership to 'maintain professional competence and knowledge'.[7] A template form is used to set objectives and to log CPD undertaken.

A 2018 report produced by the UK Landscape Institute revealed the skills landscape architects felt they needed. Based on a survey of more than 800 landscape professionals the survey listed 10 areas of expertise.[8] By far the largest area was writing plans, reports and evaluations, with 82% of respondents feeling this was a required skill.

The CPD requirement of our professional body should be the minimum standard we aspire to – throughout our careers we should aim to improve our skills and challenge our assumptions. Recognising the projects that weren't successful can provide the most valuable lessons. Human nature can make us reluctant to change, sometimes described as status quo bias, seeing that change as a loss rather than a gain. Sticking with what we know can be a sensible approach but it can also stifle improvement, creativity and innovation.

TOP PROFESSIONAL SKILLS REQUIRED BY LANDSCAPE PROFESSIONALS

Facing challenges

Attracting staff in both public sector and private practice remains a challenge – particularly for smaller firms with smaller budgets – as does retaining them in the public sector. Money is the key concern for both private and public sector respondents; for the private sector, fee levels and profitability were cited by 67% of respondents. In the public sector, concerns about accessing funding (62%) dominate. Uncertainty over budgets and priorities for private practice goes hand in hand with a lack of recognition among potential clients (48%), and not being involved at the right stage of the project (45%).

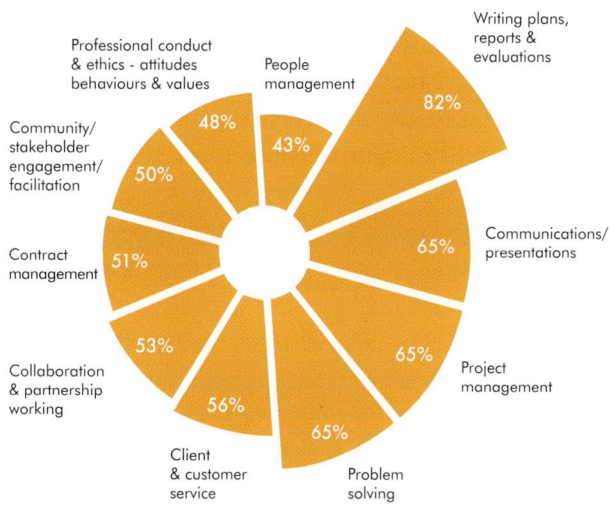

Fig 6.4 Findings from The Future State of Landscape Architecture Report, The Landscape Institute, 2018

THE CLIENT

This is likely to be the point the client has been focusing on – the moment they finally get to use the site. For us the preceding stages are our main work, but for many clients those stages are a means to get to this stage. Many of the topics relevant to the landscape architect are also relevant to the client, such as the need to look at lessons learnt, but if they are retaining ownership of the site they are also now responsible for implementing the maintenance plan.

Understanding the need for maintenance

As mentioned earlier, landscape projects need an appropriate maintenance and management plan that the client supports and will undertake. We need to ensure our clients understand that managing a landscape scheme is an ongoing process, not just something you promise to carry out to gain consent from statutory bodies, and that plans may need to be reviewed as the site establishes. Landscape schemes have so many potential variables, such as adverse weather conditions, or areas of poor soil quality that were difficult to identify, that it is almost inevitable that follow-up work such as replacing failed plant stock will be required.

The potential life span of landscape projects can make it difficult for irregular tasks to be managed, as staff move on or older projects get forgotten. We can encourage clients to set up systems so that long-term care is not overlooked, such as adding tasks to a job role rather than relying on the interest of an individual who may leave, but over long timescales finding a way to ensure maintenance tasks are reviewed can be a challenge.

We will all have seen trees where the tree ties have not been removed and the bark has been damaged, or trees constricted by a tree shelter that should have been removed. Those are relatively short-term maintenance tasks but they get overlooked. Longer term this is even more of an issue to address – how do you set up a system so that in 20 years someone gives an instruction for an area of woodland to be thinned? A well-written management and maintenance plan that the client understands and will pay for is the starting point but this only works if people know it is there.

Evaluate

For some clients there will be a requirement to evaluate the project, perhaps to report to external funders. For public-sector projects evidence is needed to justify expenditure so evidence from previous schemes may be used to demonstrate the value of planned projects.

At the very least a client should want to review whether the aims of the brief were met and whether wider objectives were achieved. Writing a brief is a leap of faith, as our client is predicting how a project may run with no idea of how outside factors may impact on the project. However, the brief should always be the basis of the project and if the final result has strayed significantly from that originally envisaged the changes are worth exploring.

Many public-sector projects in the UK have a requirement to fully evaluate projects on completion. HM Treasury produces two documents that provide guidance to central government on monitoring and evaluation. *The Green Book – Central Government Guidance on Appraisal and Evaluation 2018*[9] covers broad issues such as policy and contract appraisal as well as evaluation, and *The Magenta Book – Guidance for Evaluation 2011*[10] which focuses primarily on policy evaluation but includes guidance on the steps common to most evaluation processes.

The Green Book includes a useful definition of evaluation:

'Evaluation is the systematic assessment of an intervention's design, implementation and outcomes. It tests:

- if an intervention is working or worked
- if the costs and benefits were as anticipated
- whether it had any other consequences
- whether the consequences were anticipated
- how well it was implemented'

Some clients will want to look at a wide range of aspects, such as environmental, social or economic impact in both the long and short term. As mentioned earlier, as part of the Living Building Challenge projects have to run for a year and demonstrate that they met predicted targets before they are given full certification.

Considering lessons learnt and having an effective system shared with those who might benefit is a way for clients to improve their working practices, to prevent mistakes being repeated but also to share good practice. However, there has to be a use for any findings otherwise it is a meaningless process.

NATIONAL LOTTERY HERITAGE FUND PROJECT EVALUATION

Whilst the construction sector has been slow to include evaluation as a recognised workstage, many charitable funders have long made it a requirement of their grant-giving process. For the UK National Lottery Heritage Fund, a grant-giving organisation funded through the sale of National Lottery tickets, evaluation has been central to many landscape projects, including those funded under their Landscape Partnership grant programme.

For these multi-million-pound landscape scale schemes the evaluation costs are part of the grant, and the final grant payment is not made until a full evaluation report is submitted. Grantees must build evaluation into their project from inception, collecting baseline data at the start of the project and writing an evaluation plan. The guidance documentation is a useful reference – it asks grantees to consider if their project has 'met its objectives, as well as how effective, efficient and sustainable it was'.

Fig 6.5 The New Forest, part of a National Lottery Heritage Fund Landscape Partnership Scheme, 2017, New Forest National Park

THE PROJECT

As mentioned earlier the project is now visible to those outside the project team. Even private sites, only accessed by the client and their associates, could be featured in publications or entered into competitions – all our work has the potential to be seen and judged externally. The project moves from a live project to part of our portfolio.

If the project is a high-profile site, such as a tourist attraction, it might begin to take on a life of its own, with images shared on social media and covered in the press. Other projects may be lower key, such as habitat creation or work on infrastructure projects, and only seen by maintenance teams. It is worth remembering that less-high-profile work may still have a substantial impact – planting alongside a motorway may be seen by millions of road users per year, provide a large area of wildlife habitat and help offset the impact of traffic noise and pollution but is unlikely to gain coverage in the design media or win awards. One of the difficulties for landscape architects promoting their work is that some of our best work appears as if there has been no intervention at all. A skilfully disguised flood defence structure, a new area of woodland or a restored river channel can all be part of our work but go unrecognised.

Future changes in use

When a site is designed it isn't always possible to predict how the site will be used – trends such as skateboarding, outdoor yoga or parkour come and go, changing how a site might be used. Longer term the

Fig 6.6 Kidlington Flood Alleviation Scheme – hidden flood defences integrated into speed hump and base of Cotswold stone wall to minimise impact; Thirlwall Associates as part of project team for Environment Agency, 2009, Mill End, Kidlington, Oxfordshire

Fig 6.7 Roadside planting is unlikely to gain media coverage but it can enhance a road scheme, 2019, Didcot, Oxfordshire

usage can evolve due to numerous reasons, such as a new attraction nearby increasing visitor numbers or a thoroughfare being rerouted and decreasing footfall.

One area for landscape architects to consider is the changes in car technology and usage. The move to electric vehicles is likely to create demand for more charging points in car parks. In the longer term the use of autonomous vehicles may transform the way our towns and cities look. If cars can park themselves the area for parking reduces as there is no need to allow for opening doors, and they can park in multi-row layouts with vehicles stacked behind each other.

A 2018 study by researchers at the University of Toronto suggests that parking for autonomous vehicles will on average decrease the need for parking space by 62%.[11] Patterns of use may also change, with a model closer to that of public-hire bicycles where vehicles can be booked and used with no need for ownership, and paid for based on usage.[12]

Low-carbon site maintenance

For most landscape schemes the majority of the carbon impact of the site is likely to be during maintenance. Transporting materials long distances to site, using materials with a high energy need to create, the use of machinery on site and even staff transport all have an environmental impact but this can be over-shadowed by the year-in-year-out maintenance.

It is worth exploring the environmental impact of landscape maintenance plans – it seems counter-productive to work hard to create environmentally positive projects only to pollute as we maintain them. The traditional horticultural practices for lawn maintenance alone are not compatible with a sustainable approach – regular mowing with fossil fuel powered machinery, fertilisers, herbicides, irrigation and the creation of a low habitat value monoculture.

Choosing the correct grass seed mix for the conditions can greatly reduce the need for maintenance, including the disposal of clippings. Given the prevalence of grassed areas in our work, and in urban areas, making more sustainable choices for something as simple as lawns could have a wide-ranging impact.

In the excellent Freakonomics podcast 'How Stupid is Our Obsession with Lawns?' landscape architect Alan Turner points out that 'Grass is cheap. Grass is the cheapest ground cover you can install. The problem with grass is that it's also the most expensive ground cover to maintain.'[13]

Other issues to consider include:
- using low energy or solar lighting
- using rainwater for irrigation
- selecting species that don't require irrigation under normal conditions
- using biofuels in chainsaws and mowers or Zero Emission Equipment (ZEE)
- using biological controls
- using hand pulling or flamethrowers rather than herbicides for weed control
- specifying hard landscape items that can be repaired or small sections replaced rather than having to dispose of, an item due to minor damage
- looking at how items will be disposed of the circular representation of the workstages should act as a prompt as to how materials could be

SANTAMOUR'S RULE

Named after Dr Frank Santamour, Research Geneticist at the US National Arboretum who first proposed the concept in his paper 'Trees for Urban Planting: Diversity, Uniformity, and Common Sense' Santamour's rule states that:

'Urban foresters and municipal arborists should use the following guidelines for tree diversity within their areas of jurisdiction:

(1) plant no more than 10% of any species,

(2) no more than 20% of any genus, and

(3) no more than 30% of any family.

Strips or blocks of uniformity (species, cultivars, or clones of proven adaptability) should be scattered throughout the city to achieve spatial as well as biological diversity.'[14]

After the widespread tree loss caused in the US Dutch elm disease and chestnut blight Santamour recognised that a 'broader diversity of trees is needed in our urban landscapes to guard against the possibility of large-scale devastation by both native and introduced insect and disease pest.' The concept is hard to achieve at a small scale, but can be achieved over a wider area such as a housing development or a park.

The landscape team working for the city of Copenhagen in Denmark now follow this rule – when new street trees are planned the percentage of that tree in the immediate area is assessed so no one species exceeds 10%.[15] They also vary the species used along each street rather than the traditional two rows of identical trees. In a way this simply mimics biodiversity, nature's way of surviving rapid environmental change.

Fig 6.8 A mix of trees makes a site more resilient to disease, Basildon Park, 2018, Berkshire

Santamour sanctioned against the planting of monocultures, also noting that 'the monoculture of Homo sapiens as the prevailing intelligent life form on planet Earth has been responsible for disasters of far greater magnitude than Dutch elm disease'.[16]

FAMILY – 30%	GENUS – 20%	SPECIES – 10%	COMMON NAME/S
Fagacae	*Quercus*	*Quercus robur*	English oak, pendunculate oak
Fagacae	*Fagus*	*Fagus sylvatica*	European beech, common beech
Rosacae	*Crataegus*	*Crataegus monogyna*	Common hawthorn
Rosacae	*Sorbus*	*Sorbus intermedia*	Whitebeam
Pinaceae	*Cedrus*	*Cedrus libani*	Cedar of Lebanon
Pinaceae	*Larix*	*Larix decidua*	European larch

Table 6.2 Tree varieties

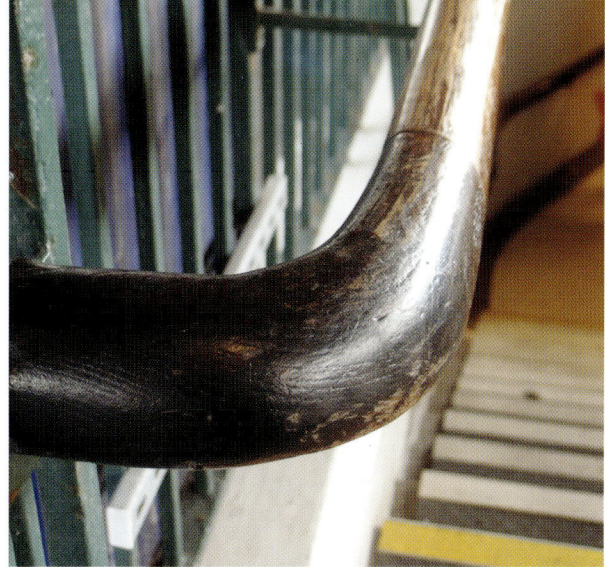

Fig 6.9 Wear and tear can be a useful indicator of popularity – handrail polished through years of intensive use, 2018, Bristol Temple Meads Station, Bristol

Fig 6.10 Area under hollowed out tree worn through play, Stowe Gardens, 2014, Buckinghamshire

reused if a site is redeveloped. The Living Building Challenge favours the use of materials where the value increases over the life of a scheme, such as weathered stone, rather than the traditional approach of items being seen as degrading over the project life span.

Disease

As mentioned earlier in this chapter disease is not something that can be easily prevented. However there are steps that can be taken to reduce the risk of disease and also to reduce the impact if disease is found.

Evaluation

We can't predict the way our projects will be used in the future and any changes that might need to be made – these could be anything from changes in legislation that means certain materials are no longer considered safe or cultural changes such as a demand for WiFi infrastructure in open spaces – but to identify the changes required we need to understand how the site is being used and if that conflicts in any way with the original design.

Techniques that can used to evaluate a site include:

- comment cards – simple cards left on site to provide general feedback
- complaints – the level or change in level of complaints can be an indicator of user satisfaction
- mystery shoppers – arrange for a mystery person to visit the site and give independent feedback

MONITORED LANDSCAPES – FLOOD NETWORK

When network architect Ben Ward's Oxford home was flooded with minimal warning he decided to investigate better ways to receive warnings of rising water levels. Though Oxford has more river level sensors than many cities, the real-time information collected failed to provide Ben with adequate warning to prepare his home.

Using his skills in wireless technologies and the Internet of Things Ben developed a prototype low-cost, battery-powered flood sensor that was small enough to be placed in discreet locations, such as under bridges or fixed to fences. A fraction of the size and cost of the telemetry used by the Environment Agency, the government organisation tasked with flood warning and prevention, the devices connect wirelessly to a gateway that sends the data via the Internet.[17] A web map then visualises the water levels at all the sensed locations, combining data from Flood Network sensors with Environment Agency data.

Individuals can use the information to make better decisions during floods, including receiving alerts, but it also provides useful benchmark data for those working in water management. Readings are taken every 15 minutes, and by using long-range low-power

wireless technology called LoRaWAN monitors can connect over several kilometres without using the mobile network. The system is routed through any gateway connected to The Things Network, an open, free worldwide Internet of Things sensor network anyone can held build.

As the size and cost of sensors reduces they can be used to:

- assess soil moisture levels in planting beds, triggering irrigation or a request for a watering visit or reporting waterlogging
- monitor plant health using infrared imaging – this is covered in more detail later in this chapter
- detect movement – if an item that shouldn't move moves, such as a

lighting column or a bridge, an alert can be sent

- manage smart lighting systems using light and weather sensors to provide adaptive street lighting.

Other sensor types include rainfall, humidity, water flow, water quality and air quality.

Flood Network is a useful example for landscape architects as to how technology can help us monitor our work, both prior to design to provide accurate baseline data and in use to assess performance. It also demonstrates the value of creating a shared and open data network, and how the Internet of Things could transform the way we work.

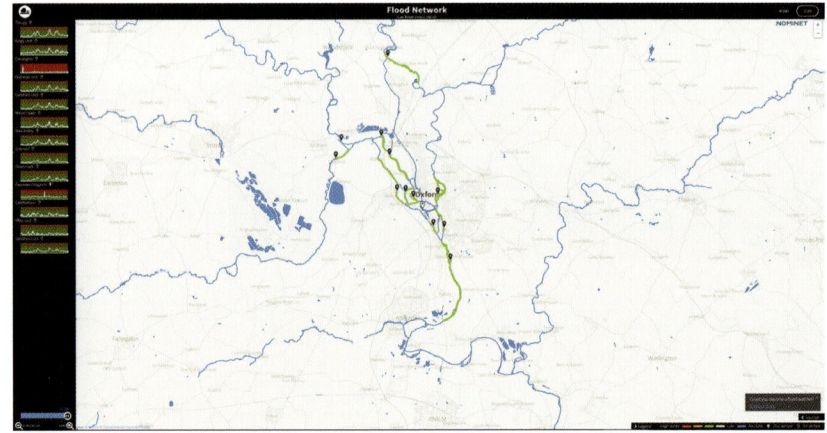

Fig 6.11 Flood Network online map, 2019

- observation – in the same way as observing a site prior to design, observing the site in use can be insightful
- online questionnaires – examples include Google Forms, Microsoft Forms, Survey Monkey
- paper questionnaires – simple technology shouldn't be overlooked. Entry to a prize draw for completed responses can increase response rates
- photographs – use fixed-point photography to record the development of a landscape, or take images to record key milestones
- previews and media coverage – monitor and review discussions on blogs, forums, review sites, social media and Google. There are tools available to analyse social media sentiment
- trialling new ideas in mock-up on site to test – a temporary fence, seating or planter to test changes, observe the impact and ask for feedback
- user groups – ask local interest groups, such as those representing parents of young children or those with a health issue, to assess the site
- visitor tracking – either with people counters, CCTV, site observations, tracking WiFi or mobile phone analytics, or using apps such as the CommonSpace app produced by Sidewalk Labs, part of Google's parent company Alphabet (in beta form and only available in the US and Canada at the time of writing but potentially a useful tool for landscape architects)
- vox pops – short recorded interviews done in situ
- wear and tear – shortcuts as people create new desire lines, perhaps to new features that weren't included in the original design, or other signs of wear and tear can show popularity.

THE PROJECT TEAM

'To avoid criticism say nothing, do nothing, and be nothing.'
Elbert Hubbard, American writer and philosopher[18]

As the project reaches a close it is a useful point for the project team to assess their individual and team performance. Was the brief met? Did the team work well together or did personality clashes mar progress? What would be done differently if we repeated the project? Knowing what we know now, would we have taken the commission?

Maintain

Processes need to be in place to ensure that the landscape scheme is maintained and managed. Maintenance is the day-to-day care of the site, ensuring that it stays as close as possible to the original design, whereas management is the long-term planning of the site, making changes and adapting the maintenance regime so that the objectives for the site are met. Management might be the responsibility of an individual who decides on a day-to-day basis if the maintenance is correct. For other sites a group might be responsible for management decisions at a more strategic level, such as a stakeholder group or a board. A management plan should show how this decision making is done and who is involved, and how the maintenance plan will be updated to reflect changes in management.

If we work in-house for the client, such as a public-sector organisation, we might continue to be part of the ongoing management and maintenance process long after handover. If we were only commissioned to the end of the construction stage, or for a short maintenance period, the project team may leave the project long before it is fully established.

Even when the project team is no longer directly appointed to help care for the site, they often retain a vested interest in the management and maintenance of a site if they are publicly associated with it, perhaps as case studies or in publications.

We may also have a legal interest in the site – under English and Welsh law professionals can be liable for negligence up to 15 years after the 'act of negligence' occurs.[19]

Evaluate

It can be helpful to evaluate a project as a team as each profession will see an issue from a different viewpoint depending on the focus of their discipline – perhaps structural issues may have been resolved but that could have been to the detriment of amenity or habitat value, or material choices were a compromise between cost and appearance.

As mentioned earlier, some clients may have a formal evaluation process that has been integral to all the workstages, with targets set from the outset. Public sector or charitably funded projects may have tracked volunteer involvement, area of habitat created or restored or other factors in addition to adherence to budget or timescale.

Sharing lessons learnt and acting on these lessons is not just about small refinements in working practice.

The fatal fire in Grenfell Tower in London in June 2017, and the subsequent investigations into the causes, highlight a culture within our sector of allowing lessons to be overlooked. The 2018 BBC TV programme 'The Fires that Foretold Grenfell', produced in association with the Open University, explored the story of five fires that provide clear lessons about fire risk in tower blocks, and provide a valuable reminder for all those working in construction about the potential risks of our decisions.

Identifying lessons learnt

Lessons learnt need to be identified and shared so they can be used to inform future projects. For large infrastructure projects in the UK there is often a requirement to share lessons learnt, with a 'learning legacy' section of the project website where good practice and innovation can be shared. These can be very detailed – the learning legacy site for the London 2012 Olympic Park, now accessible via the UK National Archives website, includes two-page PDF micro reports covering topics including sustainable material use in paving and seating and restoring the Olympic Park waterways.[20]

Any documents produced to share lessons learnt must have a final use otherwise it is a wasted process. It might not be appropriate to publicly share all lessons learnt but at the very least there should be a mechanism to ensure knowledge at an organisational level so innovation isn't lost or mistakes repeated. This could be as an in-house Wiki site – open, easy-to-edit collaborative webpages that can be used to create searchable records. Wiki sites are a wonderful way to share and collaborate

EVALUATION – THE STOCKHOLM ENVIRONMENT INSTITUTE

Good evaluation comes from good questioning. Annemarieke de Bruin, a researcher at the Stockholm Environment Institute, uses a deceptively simple set of criteria as part of her research into the relationship between people and the environment.[21] The Stockholm Environment Institute is an international non-profit research and policy organisation that works to tackle environment and development challenges, and Annemarieke works across Europe, Australia and Canada on topics including water management and governance, tree health management and agricultural innovation. As part of her work she oversaw the development of the Institute's monitoring and development system.

When planning project evaluation staff determine 'changes they "expect to see", more ambitious changes they would "like to see", and transformative changes they would "love to see"'. Staff are also encouraged to look at attitude and passing on knowledge and to understand that the success of the process itself can equal overall success.

They are also asked to consider what the perception of the project is – has it been a success? Are there ideas for stage 2? Has there been significant change in x? (follow-up question) Why is it significant? The response to this can be positive or negative.

Staff are also encouraged to look at wider outcomes outside the original scope of the project.

- Levels of continued contact – for example, a year later do those involved still speak to any other participants or have they used the knowledge gained in other projects?
- Have there been other benefits gained, such as new areas of interest, funding sources, volunteers, etc.?
- What new relationships have been formed?

Results are shared on project pages on the Institute's website, issued as internal reports and used as the basis for reports to funders and the board.

on a topic. The Designing Buildings Wiki is supported by several UK construction organisation, and is an excellent resource for landscape architects.

Project team evaluation should adopt the ethos of alliance contracting and be undertaken 'in the spirit of mutual trust and cooperation'. Evaluation isn't about blame.

Recording for future generations

Researching a book makes you realise the value of well-managed data. I owe a debt to all those who have painstakingly archived projects and research work and made it accessible online. It is worth remembering that projects that we may not consider high profile or of wider interest may be of interest to future researchers.

We may never envisage our work being looked at again in the future, but consistently including the relevant information in all documents will mean that if it does form part of future research it can be used as a source with the correct information for citation. I have found reports by academic researchers and government agencies that have no date or no author listed. Data should include:

- title
- creator

LANDSCAPE INSTITUTE ARCHIVE

From its inception in 1929 the UK Landscape Institute archive has collected books and archives relating to the profession.[22] The Landscape Institute archive and library was established in 1967 with the aim of creating a national landscape collection. In the 1990s, as the pioneers of the profession died, their collections were acquired, helping to build a resource that includes the collections of Brenda Colvin, Sylvia Crowe and Geoffrey Jellicoe.

After a number of homes the archive is now held at the Museum of English Rural Life, part of the University of Reading. The archive includes books, journals, press cuttings, minutes, photographs and financial papers and in 2019 the archive began a project to seek out existing oral history related to landscape architecture and to create new recordings. Part of the collection is available online and provides a fascinating insight into our profession.

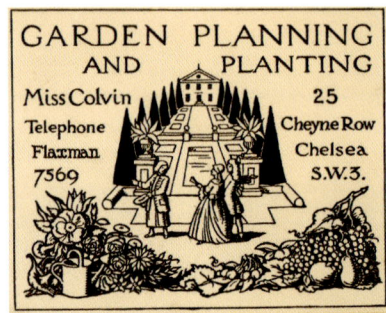

Fig 6.12 Brenda Colvin's business card – we need to remember that seemingly mundane items may be of interest to future generations, Brenda Colvin Collection at the Landscape Institute Archive

- creation date
- place of publication
- publisher/producer/distributor
- contact information
- copyright and permissions, such as Creative Commons.

With digital documents, such as PDFs or Microsoft Word files, this information can be attached to the file as metadata, rather than just in the document. This is often part of quality-assurance procedures, and is useful when categorising documents for archiving.

Opening old digital documents can be a challenge, as anyone trying to open old drawing files will be aware. The Digital Preservation Coalition produces a Digital Preservation Handbook which gives excellent guidance on the problems of digital preservation and tools to manage the process.[23] The American Library of Congress provides guidance on file formats, including those for non-photographic digital images such as architectural drawings.[24]

Taking the time to create a resilient process to archive work will ensure that work isn't lost and can be retrieved if a project is ever reinstated.

WIDER SOCIETY

For many of our projects the public is one of the main beneficiaries, either directly through access to the site or indirectly as part of the wider landscape. The work we do has potential impacts on so many areas, from local landscape character to reducing the impact of climate change, that it is only right that the public is involved in the care and evaluation of the projects we create.

Fig 6.13 Temporary repairs to paving in a busy urban setting, 2019, Cardiff

GROUNDS TIDY-UP DAYS

With limited budgets it can be difficult for schools to maintain their grounds. One solution developed by the landscape team at Hampshire County Council, based in the south of England, is to support schools to hold Grounds Tidy-Up Days.

Held once per term on a Saturday, pupils, teachers, parents, grandparents, carers and school governors are all invited into school to help complete a list of maintenance tasks set by the head teacher, whose support is central to the success of the day. Tasks can include simple manual tasks such as leaf clearing, pruning or turning compost but sometimes machinery is brought in to enable more substantial tasks. There is always a fire pit and food and drink is provided throughout the day. These cheap, easy-to-run events are social as well as practical and help develop a sense of ownership of the grounds.

Maintain

A poorly maintained site can impact on a local community, even if there is no physical public access. At the very least it can be a source of weed seed that causes extra work for those managing neighbouring sites. At worst it can be unsafe, with broken structures or poorly managed trees creating a risk to those passing by.

Failing to maintain a site to the level expected by the local community can damage the site owner's reputation. The expectation will depend on the location and prominence of the site – a busy city centre project is likely to attract more scrutiny than a rural industrial location.

Park friends groups

Independent voluntary user groups, sometimes known as friends groups, often work in partnership with local authorities and can provide useful support for public parks and open spaces, acting as a forum for discussing views and helping promote the site to the wider community. Friends groups can also help with maintenance, running clean-up days and reporting issues such as fly-tipping. Some groups are created as the result of a campaign to save a site, building on the enthusiasm and the network of interested parties created.

Working with a voluntary group can create different challenges to those encountered managing

CIC TADPOLE GARDEN VILLAGE

When British developer Crest Nicholson created the new Tadpole Garden Village north of Swindon, a large and rapidly growing town in the south-west of England, they wanted to conserve and maintain as many of the existing trees and hedges as possible within the site. Based on previous experience the company knew that investing in a higher-quality landscape made the site more desirable, as well as securing a greater profit, and that retaining the maintenance in-house would help make the 1,800 home development a place people wanted to live.

Working with local landscape practice David Jarvis Associates a design code was developed to ensure a consistent landscape style throughout the site and significant trees were protected with Tree Preservation Orders (TPOs) from the outset. Retaining the existing mature trees and hedgerows gave the new site a sense of permanence and the landscape scheme was the first spend, rather than a last step once the homes were sold, as occurs on many developments.

A management company, set up as a community interest company or CIC, is responsible for all public areas such as play areas and sports pitches. A form of social enterprise, the CIC reinvests any money made from income such as sports pitch fees back into the community, along with any surplus from the income from the resident service charge. Residents are involved in the running of the CIC, to the extent they have created new specialist sub-groups for specific topics.

A commitment to long-term maintenance is important to Crest Nicholson, as they know that the quality of their existing sites influences potential buyers and potential employees, as well as their wider reputation within the sector. The site has gained interest from local authorities and other organisations interested in exploring the CIC model, and has one of the highest sale rates in the housing sector.

Andrew Dobson, Managing Director at Crest Nicholson Strategic Projects, recognises it isn't a model that would work for all sites – the CIC was primed with a substantial initial investment and in less affluent areas that outlay would not be financially viable, nor would the 65% open space ratio that the site achieves. Based on his experience with this and other sites Andrew views the landscape scheme as an asset, though he has had to argue the case for spending on such an intangible asset. For him investing in and maintaining a quality landscape scheme is always a leap of faith, but one based on his experience he believes is worth doing.

Fig 6.14 Play area in Town Park; David Jarvis Associates for Crest Nicholson, 2019, Tadpole Garden Village, Swindon, Wiltshire

Fig 6.15 Entrance to Town Park; David Jarvis Associates for Crest Nicholson, 2019, Tadpole Garden Village, Swindon, Wiltshire

Fig 6.16 Children playing on the climbing frames in the playground; Sheffield Corporation City Architect's Department, 1963, Park Hill Estate, Sheffield, South Yorkshire

Fig 6.17 Children enjoying a free-to-use natural play area, Stanwick Lakes, 2014, Nene Valley, Northamptonshire

a site with paid staff or contractors, and if paid staff are required to support a group it isn't necessarily a less expensive option. However, it does encourage the community to become directly involved and support the site, in turn hopefully making it more resilient to outside threats.

Evaluate

The nature of many of our projects means they will be judged by those with no connection to the design of the site, and with no knowledge of the multitude of decisions that were taken to achieve the final design that they experience. For public sites there may be disappointment that not all the perceived needs were met, or for a project that had limited community support it might exceed expectations.

It is hard to know how our sites will be assessed through their usable lives – as discussed in Chapter 1 public attitude can suddenly change. Views on the costs or processes needed to maintain a site may change, or on the perception of the safety of public open spaces. Compare the play areas of 1970s Britain, with high slides, no safety surfaces and dangerous roundabouts, to recent playgrounds designed to natural play principles.

PUBLIC LAB

US-based Public Lab is a great example of how community volunteers can undertake in-depth evaluation using low-cost DIY techniques.[25] After the 2010 Gulf of Mexico oil disaster in 2010 three of Public Lab's would-be founders created their own 'community satellites' to monitor the spread of oil. Frustrated by the sparse information issued to the community over 100 local volunteers used helium balloons, kites and inexpensive digital cameras to take more than 100,000 images of the spill. These were stitched together using an open-source platform created by the group, and then uploaded to Google Earth so they were publicly accessible.

The success of the oil spill monitoring led to the creation of Public Lab, an online and offline community that supports environmental monitoring and research. The ethos is still based around low-cost, DIY techniques – recent projects include instructions for a Lego Spectrometer to test water samples and techniques for monitoring storm water run-off.

One project of interest to landscape architects is the Infragram project, supported by Google and NASA.

Near-infrared photography is used in agriculture to assess plant growth and manage soils, but has previously required expensive sensors on aircraft or satellites. Public Lab have developed an alternative using either minor adaptations to inexpensive digital cameras or their low-cost kits. Using this DIY technique volunteers can monitor the environment.

Public Lab projects have used their open-source, community-based research techniques to monitor river restoration projects, invasive aquatic plants and open landfill sites, as well as providing mapping for a Lebanese refugee camp.

Fig 6.18 Sometimes the best form of evaluation is use – popular play area at London Wetland Centre, Wildfowl and Wetlands Trust 2018, Barnes, London

Landscape projects rarely have an end point. Our work as landscape architects is to model a small area of the surface of the earth to meet a particular need in a particular timeframe. A scheme may be perfect at the moment it is created but over time become redundant and need updating. Landscape architecture is an evolution rather than a solution – unlike buildings no landscape is ever 'finished'. As a project completes on site we should be using the lessons learnt to inform our ideas for future projects. Sites change moment by moment, as plants grow and materials weather. By evaluating our work and sharing the lessons learnt we can only improve our role in this process as well as better fulfilling the brief set by our clients.

Case Study 6.1
CITY OF LYON
SUSTAINABLE LANDSCAPE
PROGRAMME

Title
Lyon – Sustainable Landscape Programme

Client	Location
City of Lyon, France	City of Lyon, France

Project Period	Implementation Period
2003-2006	2006 to present

Type of Scheme
Public realm – landscape management and maintenance

Project Value	Landscape Architect
unknown	In-house team

Owner
City of Lyon

Project Team
Speciality consultants - Howard Wood

Contractor
In-house team

In July 2013 the University of Bristol held a Sustainable Landscapes for the Future: Research, Design and Management workshop, with a series of speakers who triggered an interest in a number of topics that in turn became the basis of this book – soils, carbon sequestration, carbon grasses, direct sown planting and urban ecology.[26]

One talk that particularly caught my interest was by Howard Wood, environmental and sustainability consultant, director of Landscape and Environmental Services Ltd, about a topic I felt should be better known.

In the early 2000s Howard worked with the Parks Department in Lyon, a French city with 395 hectares of parks and open spaces, to develop a Sustainable Landscape Programme, a three-year process that was implemented in 2003.

Fig 6.1.0 (Pages 207) View of Lyon showing the historic city centre, 2019, Lyon

Fig 6.1.1 (Page 208) Urban meadow along roadside, 2008, Lyon

Fig 6.1.2 Leaf mould heaps at edge of woodland, 2003, Lyon

The move to a more sustainable approach to landscape maintenance and management was prompted by an ever-increasing budget and local politicians demanding a more environmentally aware approach.[27] The city's extensive parks network required a staff of more than 300 gardeners, and included a botanic garden, a plant nursery and a zoo. This was all managed within the historic city, founded by the Romans and in part covered by a UNESCO World Heritage designation.

Howard defines a fully sustainable landscape as a situation 'where natural cycles perpetuate without man's influence, without inputs (fertiliser, pesticides), without exports (unwanted biomass) and without maintenance (grass cutting, pruning, weeding). Everything is recycled naturally – usually where it falls.'

A programme was devised with three distinct themes:

- environment – each measure should have a favourable environmental effect
- training – the workforce needed to adopt new procedures and techniques
- financial – costs were monitored.

The themes led to the need to find practical solutions to day-to-day issues within the department, such as:

- reducing the green waste budget, which cost €250,000 for 3,000 tonnes of waste per year
- reducing the time spent cutting grass
- finding techniques for less labour-intensive floral displays
- achieveing the political aim of 0% pesticide and herbicide use.

Fig 6.1.3 Hand held butane flame used for weed control, 2004, Lyon

Over a year trial projects were run on different sites, before being adopted across the city. Green waste was processed through co-composting (mixing organic waste with faecal sludge), earthworm culture and mulching on small sites around the city.

Leaves were spread out on the woodland floor at different depths and the breakdown and absorption of organic matter measured, as well as a cost analysis. The results showed that this process saved €133,800 per year in green-waste management costs.

Fig 6.1.4 (Page 211) Parc du Vallon de la Duchère, Ilex puysage + urbanisme landsape architects 2014, Lyon

The Lyonnaise politicians set a target of 0% pesticide and herbicide use within 5 years, so biological controls were introduced, such as insect pheromones and predator-release programmes. Three thermal weed-killing techniques were trialled – an aquacide system that uses hot water, an organic biodegradable hot foam process called Waipuna™, and a butane flame. As a result herbicide use was reduced by 70% in 2 years and by more than 90% in 5 years. The cost was €32,000 more than using chemical techniques, but this was offset against savings elsewhere.

In addition grass-cutting areas, height and frequency were modified, with differential mowing in some locations, which saved €45,700.

Lower-maintenance planting styles were adopted, with a greater tolerance of weeds in areas subject to less scrutiny, such as roadsides. Over 70% of the labour-intensive traditional bedding was replaced with areas of urban meadow, saving a further €48,500.

Care was taken to communicate the changes to the public, to reassure them that the landscape was still being cared for and to alleviate complaints, with monthly updates posted on park noticeboards.

Fig 6.1.5 Differential mowing along the edge of a park path, 2008, Lyon

Fig 6.1.6 Roadside planting – a simpler style and lower level of maintenance, 2008, Lyon

The community were also involved in the work itself, with typically 10% of contract values being spent on training and employment for disadvantaged members of the community.

After three years the results of the Sustainable Landscape Programme were positive:

- increased environmental benefits, saving 20,000km of HGV transport in the city, green waste recycled locally and a 90% reduction in herbicide use
- increased local community employment
- 13.85% reduction in the park department budget, with the potential for further savings of up to 20–25%.

Since the completion of the programme the city has continued to follow the principles. The use of pesticides ceased in 2008, irrigation management has been improved and sustainable procurement implemented, requiring FSC timber, non-polluting materials and a preference for short supply chains, and the city is working to reduce energy use. Since 2010 the city has seen a 21% reduction in greenhouse gas emissions and the Lyon – Ville Équitable et Durable initiative (Lyon – a fair and sustainable city) is now central to the city's policies.

Fig 6.1.7 Simpler planting styles to reduce maintenance costs, 2008, Lyon

Fig 6.1.8 Traditional formal, labour intensive bedding, 2008, Lyon

The European Region of the International Federation of Landscape Architects Code of Ethics and Professional Standards [1]

PROFESSIONAL ATTITUDES

Standard 1. To promote the highest standard of professional services, and conduct professional duties with honesty and integrity, having regard to the interest of those who may be reasonably expected to use or enjoy the products of their work.

Standard 2. To support continuing professional development.

Standard 3. To uphold the reputation and dignity of the profession, IFLA/IFLA EUROPE and their own professional organisations, respecting the resolutions of the respective General Assemblies, Executive Councils, Boards, Committees and Working Groups, as well as their external communications events and social networks.

Standard 4. To actively and positively promote the standards set out in this Code of Ethics and Professional Conduct.

Standard 5. To be fully acquainted with the Statutes and Regulations of IFLA EUROPE and their own professional association(s), and be willing to co-operate – in any possible way and with the due dedication and independence of judgment – in achieving the aims and objectives of their respective Strategic and associated Action Plan(s).

Standard 6. To observe all laws and regulations related to the professional activities of landscape architecture in their respective countries.

Standard 7. To act at all times with integrity and avoid any action or situations which are inconsistent with their professional obligations.

Standard 8. To be fair and impartial in all dealings with clients' contractors, and at any level of arbitration and project evaluation.

Standard 9. To make full disclosure to the client or employer of any financial or other interest relevant to the service or project. In particular, IFLA EUROPE members who have economic interests in construction companies or suppliers of the proposed works shall be obliged to inform their clients and obtain the corresponding authorisations.

Standard 10. To refuse to take charge of tasks or projects in conflict of rights/interests or in conditions of incompatibility, especially in case they are state employees or hold any positions at public bodies, as established by the current civil legislation of the involved country(ies).

Standard 11. To refuse to accept equivocal positions that could jeopardise their righteousness or independence in properly carrying out the profession.

Standard 12. To avoid participating in competitions for which they accepted to serve as members of the Panel of Judges or helped define terms and requirements, or where there are anyhow involved people with whom they have family or business relationships.

Standard 13. To undertake public service in local governance and environment to improve public appreciation and understanding of the profession and environmental systems.

PROFESSIONAL COMPETENCES

Standard 14. To undertake only professional work for which they are able to provide proper professional and technical competence and resources.

Standard 15. To maintain qualified professional competence in areas relevant to their own professional work, and carry out their profession work with care, conscientiously and with proper regard to the specific technical and professional standards.

PROFESSIONAL RELATIONSHIPS

Standard 16. To organise and manage their professional work responsibly and with integrity, having constant regard to the interests of their clients.

Standard 17. To promote their professional services in a truthful and responsible manner, without misleading or deceptive claims discreditable to the profession or the work of other professionals.

Standard 18. To uphold maximum respect for the colleagues of their own and any other member association, its representatives and boards, avoiding making statements personally offensive to their peers or to the profession.

Standard 19. To provide, in a timely fashion, all information, explanations, documents or reports they might be asked for by IFLA EUROPE or their own professional association(s).

Standard 20. To promote the exchange, discussion and debate in IFLA EUROPE – live or by means of its social networks – in a truthful and responsible manner, without deceptive claims to, or bringing discredit on, or insulting the IFLA/IFLA EUROPE organisations, officers, member associations, representatives and members of any membership category, as well as any other professional whether working or not as landscape architect.

Standard 21. To inform IFLA EUROPE and the respective national association(s) of any breach of professional duties or misconduct they might be aware of.

Standard 22. To ensure local culture and place are recognised by working in conjunction with a local colleague when undertaking work in a foreign country.

Standard 23. To act in support of other landscape architects, colleagues and partners in their own and other disciplines. Where another landscape architect is known to have undertaken work for which the member is approached by a client, to notify the professional colleague before accepting such commission.

Standard 24. To provide educational and training support to less experienced members or students of the profession over whom they have a professional or employment responsibility.

Standard 25. To manage their personal and professional finances prudently and to preserve the security of monies entrusted to their care in the course of practice or business.

Standard 26. To respect the fee regulations of the profession in countries where such regulations exist.

Standard 27. To participate only in planning or design competitions which are in accordance with the approved competition principles and guidelines of IFLA/IFLA EUROPE, or of IFLA/IFLA EUROPE member organisation in the respective country.

Standard 28. To have an adequate and appropriate Professional Indemnity Insurance.

Standard 29. To deal with any complaints concerning their professional work or practice promptly and appropriately.

LANDSCAPE AND ENVIRONMENT

Standard 30. To recognize and protect the cultural and historical context and the ecosystem to which the landscape belongs when generating design, planning and management proposals.

Standard 31. To develop, use and specify materials, products and processes which exemplify the principles of sustainable management and landscape regeneration.

Standard 32. To advocate values that support human health, environmental protection and biodiversity.

Approved by the General Assembly at its meeting held in Oslo, Norway on October 19th, 2014

Boyd, D, & E Chinyio, *Understanding the Construction Client.* Oxford; Malden, MA, Blackwell, 2006 – one of the few books on the client relationship

Crates, E, *Building a Fairer System: Tackling Modern Slavery in Construction Supply Chains.* The Chartered Institute of Building (CIOB), 2016

Crawford, M, *Creating a Forest Garden: Working with Nature to Grow Edible Crops.* Totnes, Devon, Green Books, 2012 – a comprehensive work useful for designing edible landscapes with many topics relevant to landscape architecture

Farmer, M, *'The Farmer Review of the UK Construction Labour Model: Modernise or Die – Time to Decide the Industry's Future'.* Construction Leadership Council (CLC) – worth reading in full to understand how the construction industry needs to change

Hitchmough, J, Sowing *Beauty: Designing Flowering Meadows from Seed.* Portland, USA, Timber Press, 2017 – a stunning book explaining James's planting concept

Integrated Security: *A Public Realm Design Guide for Hostile Vehicle Mitigation* – Second Edition. Centre for the Protection of National Infrastructure, 2014, https://www.cpni.gov.uk/system/files/documents/40/20/Integrated%20Security%20Guide.pdf – this important guide won the Landscape Institute Awards 2011 Research Award

Landscape Institute, ed., BIM for Landscape. London; New York, Routledge, Taylor & Francis Group, 2016 – a well-written and informative book on BIM specific to our profession

Moggridge, H, *Slow Growth: On the Art of Landscape Architecture.* London, Unicorn, 2017 – one of my favourite books on landscape architecture; Hal's insight into our profession is a great reminder of the value of our art

Rosling, H, O Rosling, & AR Rönnlund, *Factfulness: Ten Reasons we're Wrong about the World – and Why Things are Better than you Think.* London, Sceptre, 2018 – a wonderfully inspiring book by the late Hans Rosling. A good primer for interpreting data – after reading it you'll start to spot the tricks used by politicians

Peters, S, *The Chimp Paradox: The Mind Management Programme for Confidence, Success and Happiness.* London, Vermilion, 2012 – one of the best books I've read on behaviour, and useful for understanding teams

REFERENCES

Chapter 1

1 Based on the Landscape Institute Digital Plan of Works for Landscape – Release 1.0, created by Anna Dekker and the LI BIM Working Group for The Landscape Institute, 2017 <https://www.landscapeinstitute.org/technical/bim-working-group/li-digital-plan-of-works-for-landscape/> [accessed 18 July 2019 and licensed under CC BY 4.0].

2 R Cohen, C Bavishi & A Rozanski, 'Purpose in Life and Its Relationship to All-Cause Mortality and Cardiovascular Events: A Meta-Analysis', in *Psychosomatic Medicine*, vol. 78, 2016, 122–133.

3 'The Ten Principles | UN Global Compact', <https://www.unglobalcompact.org/what-is-gc/mission/principles> [accessed 12 October 2018].

4 LIVING BUILDING CHALLENGE 4.0 SM A Visionary Path to a Regenerative Future, International Living Future Institute, <https://living-future.org/lbc/> 2019.

5 5 'Certification | SITES', <http://www.sustainablesites.org/certification-guide> [accessed 26 July 2019].

6 M Virtanen et al., 'Long working hours, anthropometry, lung function, blood pressure and blood-based biomarkers: cross-sectional findings from the CONSTANCES study', in *Journal of Epidemiology and Community Health*, 2018, jech-2018.

7 E van der Helm, N Gujar & MP Walker, 'Sleep Deprivation Impairs the Accurate Recognition of Human Emotions', in *Sleep*, vol. 33, 2010, 335–342.

8 'Driver fatigue – Brake the road safety charity', <http://www.brake.org.uk/facts-resources/15-facts/485-driver-tiredness> [accessed 25 October 2018].

9 MM Mitler et al., 'Catastrophes, Sleep, and Public Policy: Consensus Report', in *Sleep*, vol. 11, 1988, 100–109.

10 'The Future State of Landscape Architecture – Embracing the opportunity', The Landscape Institute, 2018.

11 T Amabile, 'How to Kill Creativity', in *Harvard Business Review*, 1998, <https://hbr.org/1998/09/how-to-kill-creativity> [accessed 29 November 2018].

12 'Suicide by occupation, England – Office for National Statistics', <https://www.ons.gov.uk/peoplepopulationandcommunity/birthsdeathsandmarriages/deaths/articles/suicidebyoccupation/england2011to2015> [accessed 28 November 2018].

13 'Suicides in Great Britain – Office for National Statistics', <https://www.ons.gov.uk/peoplepopulationandcommunity/birthsdeathsandmarriages/deaths/bulletins/suicidesintheunitedkingdom/2016registration> [accessed 28 November 2018].

14 'Men's Mental Health and work: The Case for a Gendered Approach', The Work Foundation, 2018.

15 'Action Plan · IFLA World', <http://iflaonline.org/about/structure-governance/regulatory-documents/action-plan/> [accessed 28 February 2018].

16 'IFLA Europe Code of Ethics and Professional Conduct', IFLA Europe, 2014.

17 'The Landscape Institute Code of Standards of Conduct and Practice for Landscape Professionals', The Landscape Institute, 2012.

18 'American Society of Landscape Architects Code of Professional Ethics', 2019, <https://www.asla.org/uploadedFiles/CMS/About__Join/Leadership/Leadership_Handbook/Ethics/ASLA%20Code%20of%20Professional%20Ethics.4.2.19.pdf>.

19 'ASLA Code of Environmental Ethics | asla.org', <https://www.asla.org/ContentDetail.aspx?id=4308&RMenuId=8&PageTitle=Leadership> [accessed 17 June 2019].

20 'Australian Institute of Landscape Architects Code of Professional Conduct', Australian Institute of Landscape Architects, 2005, <https://www.aila.org.au/imis_prod/

documents/AILA/Advocacy/AILA%20Policies/code-of-conduct.pdf>.

21 'Etiskt program och etiska regler', in *Sveriges Arkitekter*, 2014, <https://www.arkitekt.se/etiska-regler/> [accessed 16 May 2019].

22 'Irish Landscape Institute – Code of Ethics and Professional Conduct', 2012, <http://www.irishlandscapeinstitute.com/wp-content/uploads/2013/09/ILI_CodeofEthicsandProfessionalConduct_2012.pdf>.

23 'Construction – Construction Design and Management summary of duties', <http://www.hse.gov.uk/construction/cdm/2015/summary.htm> [accessed 17 June 2019].

24 'Quarterly Overdue Payments Report Q4 2015', <https://www.eulerhermes.co.uk/press_imported/quarterly-overdue-payments-report-q4-2015.html> [accessed 30 November 2018].

25 'Carillion declares insolvency: information for employees, creditors and suppliers', in GOV.UK, <https://www.gov.uk/government/news/carillion-declares-insolvency-information-for-employees-creditors-and-suppliers> [accessed 18 October 2018].

26 'Carillion attacked over subcontractor payments', in *Financial Times*, 2013, <https://www.ft.com/content/b90c2286-b331-11e2-b5a5-00144feabdc0> [accessed 18 October 2018].

27 'The hidden dangers of Carillion's new payment terms', <https://www.theconstructionindex.co.uk/news/view/the-hidden-dangers-of-carillions-new-payment-terms> [accessed 12 October 2018].

28 'Prompt Payment Code', <http://www.promptpaymentcode.org.uk/> [accessed 18 October 2018].

29 F Coppola, 'How Carillion Used A U.K. Government Scheme To Rip Off Its Suppliers', in *Forbes*, <https://www.forbes.com/sites/francescoppola/2018/01/30/how-carillion-used-a-u-k-government-scheme-to-rip-off-its-suppliers/> [accessed 18 October 2018].

30 Business, Energy and Industrial Strategy Committee, 'Carillion – Business, Energy and Industrial Strategy and Work and Pensions Committees – House of Commons', House of Commons, 2018, <https://publications.parliament.uk/pa/cm201719/cmselect/cmworpen/769/76905.htm#_idTextAnchor009> [accessed 15 October 2018].

31 'Carillion joint inquiry', in *UK Parliament*, <https://www.parliament.uk/business/committees/committees-a-z/commons-select/work-and-pensions-committee/inquiries/parliament-2017/carillion-inquiry-17-19/> [accessed 15 October 2018].

32 T Clark, 'Carillion's true cost only just starting to emerge', in *Construction News*, <https://www.constructionnews.co.uk/analysis/cn-briefing/carillions-true-cost-only-just-starting-to-emerge/10037101.article> [accessed 13 November 2018].

33 Ibid.

34 *THE PEAK DISTRICT (QUARRYING)* (Hansard, 7 February 1949), 1949, <http://hansard.millbanksystems.com/commons/1949/feb/07/the-peak-district-quarrying> [accessed 13 March 2018].

35 FSC- International, 'History, partnership and calculated risk in times of change for FSC', in *FSC International*, <https://ic.fsc.org/en/news-updates/id/2000> [accessed 18 June 2019].

36 FSC- International, 'The Share of Sustainable Wood: Data on FSC's Presence in Global Wood Production', in *FSC International*, <https://ic.fsc.org/en/news-updates/id/2210> [accessed 18 June 2019].

37 'Welsh Government | Well-being of Future Generations (Wales) Act 2015', <http://gov.wales/topics/people-and-communities/people/future-generations-act/?lang=en> [accessed 23 August 2017].

38 'Living Building Challenge | Living-Future.org', International Living Future Institute, 2016, <https://living-future.org/lbc/> [accessed 18 June 2019].

Chapter 2

1 Based on the Landscape Institute Digital Plan of Works for Landscape – Release 1.0, created by Anna Dekker and the LI BIM Working Group for The Landscape Institute, 2017 <https://www.landscapeinstitute.org/technical/bim-working-group/li-digital-plan-of-works-for-landscape/> [accessed 18 July 2019 and licensed under CC BY 4.0].

2 SJ Ashford, G Petriglieri & A Wrzesniewski, 'The 4 Things You Need to Thrive in the Gig Economy', *Harvard Business Review*, March–April 2018, 140–143.

3 'Getting a trip request', in *Uber*, <https://help.uber.com/h/6b2345ff-9260-4dca-aadf-687ae5bae7c2> [accessed 23 March 2018].

4 *The Landscape Consultant's Appointment,* 1998, The Landscape Institute.

5 M Farmer, 'The Farmer Review of the UK Construction Labour Model: Modernise or Die – Time to Decide the Industry's Future', Construction Leadership Council (CLC). www.constructionleadershipcouncil.co.uk, 2016, 13.

6 'Digital America: A tale of the haves and have-mores | McKinsey', <https://www.mckinsey.com/industries/high-tech/our-insights/digital-america-a-tale-of-the-haves-and-have-mores> [accessed 18 June 2019].

7 'Soil as Carbon Storehouse: New Weapon in Climate Fight?', in *Yale E360*, <https://e360.yale.edu/features/soil_as_carbon_storehouse_new_weapon_in_climate_fight> [accessed 18 June 2019].

8 SE Ward et al., 'Legacy effects of grassland management on soil carbon to depth', in *Global Change Biology*, vol. 22, 2016, 2929–2938.

9 University of Bristol, '2013: Sustainable landscapes for the future | Cabot Institute for the Environment | University of Bristol', <http://www.bristol.ac.uk/cabot/events/2013/298.html> [accessed 25 April 2019]. Stephen Alderton, DLF France – 'Low Maintenance & Carbon Sequestration: Top Green grass variety research at DLF France's research station, Les Alleuds Angers, France'.

10 RA Birdsey, 'Carbon storage for major forest types and regions in the conterminous United States', in *Forests and Global Change*, vol. 2, 1996, 1–26.

11 'CO$_2$ emissions (metric tons per capita) | Data', <https://data.worldbank.org/indicator/EN.ATM.CO2E.PC> [accessed 19 July 2019]. (UK average of 6.5 metric tonnes of CO$_2$ emissions per capita based on 2014 data.)

12 M Latham, *Constructing the Team: Final Report: Joint Review of Procurement and Contractual Arrangements in the United Kingdom Construction Industry,* London, HMSO, 1994.

13 T Öno, *Toyota Production System: Beyond Large-scale Production*, Cambridge, Mass. Productivity Press, 1988.

14 D Boyd & E Chinyio, *Understanding the Construction Client*, Oxford ; Malden, MA, Blackwell, 2006.

15 Lancelot 'Capability' Brown, 'The Account Book of Lancelot 'Capability' Brown, the great landscape gardener of Fenstanton, Hants', 1759–1788. Royal Horticultural Society Lindley Library. GB 803 CAP' on the Archives Hub website, <https://archiveshub.jisc.ac.uk/data/gb803-cap>, [accessed 15 July/2019] <https://archiveshub.jisc.ac.uk/search/archives/4a8b94c0-efee-355d-855f-b7d7e4365147>.

16 <https://www.english-heritage.org.uk/visit/places/audley-end-house-and-gardens/history/capability-brown-at-audley-end/>.

17 M Farmer, 'Modernise or Die!', presented at the CIOB & Constructing Excellence CPD Modernise or Die!, Saïd Business School, Oxford, 2018, <https://events.ciob.org/ehome/200176103>.

18 'Find open data – data.gov.uk', <https://data.gov.uk/> [accessed 18 June 2019].

19 QGIS is a free, open source, cross platform (lin/win/mac) geographical information system (GIS), QGIS, 2018, <https://github.com/qgis/QGIS> [accessed 25 April 2018].

20 'Find your nearest Maggie's Centre', in *Maggie's Centres*, <https://www.maggiescentres.org/our-centres/> [accessed 18 June 2019].

21 'Maggie's Architecture and Landscape Brief', Maggie Keswick Jencks Cancer Caring Trust (Maggie's), 9.

22 'Facts and Figures: Water, Mills & Marshes', in *Water, Mills & Marshes*, <https://watermillsandmarshes.org.uk/wmmdetail/facts-and-figures/> [accessed 19 June 2019].

Chapter 3

1 Based on the Landscape Institute Digital Plan of Works for Landscape – Release 1.0, created by Anna Dekker and the LI BIM Working Group for The Landscape Institute, 2017 <https://www.landscapeinstitute.org/technical/bim-working-group/li-digital-plan-of-works-for-landscape/> [accessed 18 July 2019 and licensed under CC BY 4.0].

2 H Moggridge, *Slow Growth: On the Art of Landscape Architecture*, London, Unicorn, 2017.

3 S King, *On Writing: A Memoir of the Craft*, London, Hodder, 2012.

4 'BS EN ISO 11091:1999 – Construction drawings. Landscape drawing practice', <https://shop.bsigroup.com/Pr

oductDetail/?pid=000000000030011259> [accessed 19 June 2019].

5 *Guidelines for Landscape and Visual Impact Assessment*, Landscape Institute & Institute of Environmental Management and Assessment (eds), Third edition, London; New York, Routledge, Taylor & Francis Group, 2013.

6 'Spatial opendata | Landscape Institute', <https://www.landscapeinstitute.org/technical-resource/spatial-opendata/> [accessed 19 June 2019].

7 G Jellicoe, Kennedy memorial lecture to the Royal Academy of Arts, 28 February 1967 – published in *Studies in Landscape Design*, London, New York, Oxford University Press, 1960.

8 S Mann & R Cadman, 'Does Being Bored Make Us More Creative?', in *Creativity Research Journal*, vol. 26, 2014, 165–173.

9 'Sight, perception and hallucinations in dementia– Factsheet 527LP', Alzheimer's Society, 2016, <https://www.alzheimers.org.uk/sites/default/files/pdf/sight_perception_and_hallucinations_in_dementia.pdf> [accessed 19 June 2019].

10 'Creating a dementia friendly environment | Age UK Norfolk', <https://www.ageuk.org.uk/norfolk/our-services/dementia-in-your-community/creating-a-dementia-friendly-environment/> [accessed 19 June 2019].

Chapter 4

1 Construction – Construction Design and Management Summary of Duties.

2 IFTTT, 'IFTTT helps your apps and devices work together', <https://ifttt.com> [accessed 19 June 2019].

3 'Common data environment CDE – Designing Buildings Wiki', <https://www.designingbuildings.co.uk/wiki/Common_data_environment_CDE> [accessed 19 June 2019].

4 What Clients think of Architects – Feedback from the 'Working with Architects' Client Survey 2016 <https://www.architecture.com/-/media/gathercontent/working-with-architects-survey/additional-documents/ribaclientsurveyfinalscreenwithoutappendixpdf.pdf> [accessed 16 July 2019].

5 'Bring your Lessons to Life with Expeditions', in Google for Education, <https://edu.google.com/products/vr-ar/expeditions/> [accessed 19 June 2019].

6 'Expeditions Pioneer Program – Google', <https://www.google.co.uk/edu/pioneer-program/> [accessed 19 June 2019].

7 'Materials Petal | Living-Future.org', in International Living Future Institute, 2016, <https://living-future.org/lpc/materials-petal/> [accessed 19 June 2019].

8 'Candidate List of substances of very high concern for Authorisation – ECHA', <https://echa.europa.eu/candidate-list-table> [accessed 4 October 2017].

9 'The Red List', in The Living Future Institute, <https://living-future.org/declare/declare-about/red-list/> [accessed 4 October 2017].

10 'Declare Products | Living-Future.org', International Living Future Institute, 2016, <https://living-future.org/declare/> [accessed 19 June 2019].

11 'Assessing the costs and benefits of reducing waste in construction – Cross-sector comparison', WRAP, <http://www.wrap.org.uk/sites/files/wrap/CBA%20Summary%20Report1.pdf>.

12 'Selling light as a service', <https://www.ellenmacarthurfoundation.org/case-studies/selling-light-as-a-service> [accessed 21 June 2018].

13 'DiSC Profile – Assessments for teams and team members', in DiSCProfile.com, <https://discprofile.com/which-disc-to-use/assessments-for-teams/> [accessed 19 June 2019].

14 D McGinn, 'What Companies Can Learn from Military Teams', *Harvard Business Review*, 2015, <https://hbr.org/2015/08/what-companies-can-learn-from-military-teams> [accessed 26 June 2018].

15 R Millar, 'Serve the Story', 2015, <https://www.ryanmillar.com/serve-the-story/> [accessed 25 June 2018].

16 S Peters, *The Chimp Paradox: The Mind Management Programme for Confidence, Success and Happiness*, London, Vermilion, 2012.

17 *BIM for Landscape*, Landscape Institute (ed), London; New York, Routledge, Taylor & Francis Group, 2016.

18 A Andreou, 'Defensive architecture: keeping poverty unseen and deflecting our guilt', *Guardian*, 18 February 2015, section Society, <http://www.theguardian.com/society/2015/feb/18/defensive-architecture-keeps-poverty-undeen-and-makes-us-more-hostile> [accessed 17 July 2018].

19 J Jacobs, *The Death and Life of Great American Cities*, Vintage Books (ed), New York, Vintage Books, 1992.

20 A Minton & J Aked, "Fortress Britain": High Security, Insecurity and the Challenge of Preventing Harm' in *The City Between Freedom and Security*, D Simpson, V Jensen & A Rubing (eds), Berlin, Boston, De Gruyter, 2017, <http://www.degruyter.com/view/books/9783035607611/9783035607611-009/9783035607611-009.xml> [accessed 23 July 2018].

21 'Conflict Minerals: Campaigners Strongly Criticise "Weak" EU Safeguards Against Conflict Minerals', <https://www.amnesty.org.uk/press-releases/conflict-minerals-campaigners-strongly-criticise-weak-eu-safeguards-against-conflict> [accessed 19 June 2019].

22 Great Britain & Home Office, Modern slavery strategy., 2014, <https://nls.ldls.org.uk/welcome.html?ark:/81055/vdc_100023703808.0x000001> [accessed 24 August 2017].

23 '21 Million People Are Now Victims of Forced Labour, ILO Says'.

24 Kanthal, 'Anti-slavery day: factsheet', in United Nations Association, <https://www.una.org.uk/anti-slavery-day-factsheet> [accessed 3 August 2018].

25 'Tide kills 18 cockle pickers', 6 February 2004, <http://news.bbc.co.uk/1/hi/england/lancashire/3464203.stm> [accessed 19 June 2019].

26 E Crates, 'Building a Fairer System: Tackling Modern Slavery in Construction Supply Chains', The Chartered Institute of Building (CIOB), 2016.

27 Ibid

28 Ibid. 43.

29 Ibid. 51.

30 'Stockholm – a city for everyone. Participation programme for people with disabilities 2011–2016', Stadsledningskontoret, 2011.

31 'Convention on the Rights of Persons with Disabilities (CRPD) | United Nations Enable', <https://www.un.org/development/desa/disabilities/convention-on-the-rights-of-persons-with-disabilities.html> [accessed 3 July 2018].

32 Lennart Klaesson, Catarina Nilsson, & Sara Malm with Pernilla Johnni, 'Stockholm – en stad för alla', *Trafikkontoret Stockholm*, 2008, 74.

33 'Stockholm – the City for Everyone: Twelve Years of the Project of Easy Access', 9.

34 Steve Maslin, in his book on *Enabling Mind Friendly Environments* (title to be agreed) 2019.

35 'Inclusive Design Strategy 2008', Olympic Delivery Authority I, 2008, 8.

36 'Accessibility – 2012 Olympics | London 2012', 2012, <https://web.archive.org/web/20120705115511/http://www.london2012.com/spectators/accessibility/index.html> [accessed 5 September 2018].

37 'Convention on the Rights of Persons with Disabilities – Articles | United Nations Enable'.

38 L Poon, 'Google Gets Serious About Mapping Wheelchair Accessibility', in CityLab, <https://www.citylab.com/life/2017/09/google-gets-serious-about-mapping-wheelchair-accessibility/539220/> [accessed 20 August 2018].

39 'Inclusive Design Standards 2013', 2013, 20, <http://www.queenelizabetholympicpark.co.uk/-/media/qeop/files/public/inclusivedesignstandardsmarch2013.ashx?la=en>.

40 R Ulrich, 'View through a window may influence recovery from surgery', in Science, vol. 224, 1984, 420–421.

41 R Ulrich, O Lundén & J Eltinge, 'Effects of exposure to nature and abstract pictures on patients recovering from heart surgery', Thirty-Third Meeting of the Society of Psychophysiological Research, Rottach-Egern, Germany., 1993.

42 ACK Lee & R Maheswaran, 'The health benefits of urban green spaces: a review of the evidence', *Journal of Public Health*, vol. 33, 2011, 212–222.

43 'Urban Green Spaces and Health', Copenhagen, World Health Organization Regional Office for Europe, 2016.

44 'Sustainable Development Goals: 17 Goals to Transform Our World', United Nations Sustainable Development, <https://www.un.org/sustainabledevelopment/> [accessed 28 August 2018].

45 'Urban Green Spaces and Health', 11.

46 PP Ekins, 'Reflections: Ecosystem Services and Sustainable Development', presented at Creating a New Prosperity: Fresh Approaches to Ecosystem Services and Human Well-being, Royal Geographical Society, London, 2009, <https://www.researchcatalogue.esrc.ac.uk/grants/RES-496-26-0045/outputs/read/47f489f8-ee7d-4589-80c9-a02358806339>.

47 'O'Dell Engineering – Land Connections – Vestibular & Proprioceptive Sensory Systems – Modesto Landscape Architecture', <http://www.odellengineering.com/informer/L_PA-Oct_10.htm> [accessed 30 August 2018].

48 GA Rook, 'Regulation of the immune system by biodiversity from the natural environment: An ecosystem service essential

to health', Proceedings of the National Academy of Sciences of the United States of America, 2013, <www.pnas.org/cgi/doi/10.1073/pnas.1313731110>.

49 'Road traffic statistics – Manual count point: 27663', <https://roadtraffic.dft.gov.uk/manualcountpoints/27663> [accessed 25 June 2019].

50 'Assessment of an Archaeological Watching Brief During the Whitehall Streetscape Improvement Project, City of Westminster', Pre-Construct Archaeology Ltd (London), <http://archaeologydataservice.ac.uk/library/browse/issue.xhtml?recordId=1117594&recordType=GreyLitSeries>.

51 Integrated Security – A Public Realm Design Guide for Hostile Vehicle Mitigation – Second Edition, Centre for the Protection of National Infrastructure, 2014, 22, <https://www.cpni.gov.uk/system/files/documents/40/20/Integrated%20Security%20Guide.pdf>.

52 The Institution of Highways & Transportation Awards 2008, the IHT/Centre for the Protection of National Infrastructure Security in the Public Realm Award submission Whitehall Streetscape Improvement Project.

53 Integrated Security: A Public Realm Design Guide for Hostile Vehicle Mitigation – Second Edition.

Chapter 5

1 Based on the Landscape Institute Digital Plan of Works for Landscape – Release 1.0, created by Anna Dekker and the LI BIM Working Group for The Landscape Institute, 2017 <https://www.landscapeinstitute.org/technical/bim-working-group/li-digital-plan-of-works-for-landscape/> [accessed 18 July 2019 and licensed under CC BY 4.0].

2 'BIM Level 2 Benefits Measurement application of PwC's BIM Level 2 Benefits Measurement Methodology to Public Sector Capital Assets', PwC, 2018, <https://www.cdbb.cam.ac.uk/Downloads/Level2/4.PwCBMMApplicationReport.pdf>.

3 A Luck & H Boyes, 'Introduction to PAS 1192-5:2015: A specification for security-minded building information modelling, digital built environments and smart asset management', British Standards Institution (BSI) and Centre for the Protection of National Infrastructure (CPNI), 2, <https://www.cpni.gov.uk/system/files/documents/18/6f/BIM-Introduction-To-PAS1192-5.pdf>.

4 'Construction Code of Practice for the Sustainable Use of Soils on Construction Sites', Department for Environment, Food and Rural Aff, 2009, 6, <https://assets.publishing.service.gov.uk/government/uploads/system/uploads/attachment_data/file/716510/pb13298-code-of-practice-090910.pdf>.

5 'Permits & Tree Protection', City of Toronto, 2017, <https://www.toronto.ca/services-payments/building-construction/tree-ravine-protection-permits/tree-protection/> [accessed 22 July 2019].

6 'Trees in Toronto', City of Toronto, 2017, <https://www.toronto.ca/services-payments/water-environment/trees/> [accessed 11 February 2019].

7 'Permit to Injure or Remove Trees', City of Toronto, 2017, <https://www.toronto.ca/services-payments/building-construction/tree-ravine-protection-permits/permit-to-injure-or-remove-trees/> [accessed 22 July 2019].

8 'BS 5837:2012 Trees in relation to design, demolition and construction. Recommendations', 19, <https://shop.bsigroup.com/ProductDetail/?pid=000000000030213642> [accessed 20 June 2019].

9 'Tree Preservation Orders and trees in conservation areas', in GOV.UK, <https://www.gov.uk/guidance/tree-preservation-orders-and-trees-in-conservation-areas> [accessed 15 March 2019].

10 RG Eccles, SC Newquist & R Schatz, 'Reputation and Its Risks', in Harvard Business Review, 2007, <https://hbr.org/2007/02/reputation-and-its-risks> [accessed 8 February 2019].

11 'What is Technical Debt? – Definition from Techopedia', in Techopedia.com, <https://www.techopedia.com/definition/27913/technical-debt> [accessed 20 June 2019].

12 Improving Infrastructure Delivery: Alliancing Best Practice in Infrastructure Delivery, HM Treasury, 2014, <https://assets.publishing.service.gov.uk/government/uploads/system/uploads/attachment_data/file/359853/Alliancing_Best_Practice.pdf>.

13 'What is alliancing?', LH Alliances, <http://lhalliances.org.uk/what-is-alliancing/> [accessed 15 February 2019].

14 'Constructing the Team (The Latham Report)', in Constructing Excellence, 2015, <http://constructingexcellence.org.uk/resources/constructing-the-team-the-latham-report/> [accessed 20 June 2019].

15 M Chui et al., 'The social economy: unlocking value and productivity through social technologies | McKinsey',

<https://www.mckinsey.com/industries/high-tech/our-insights/the-social-economy> [accessed 22 January 2019].

16 'Project Management Software – Edenvale Young Associates', <http://webserve.default.edenvaleyoung.uk0.bigv.io/blog/2015/11/project-management-software#more> [accessed 22 January 2019].

17 'BS 3936-1:1992 – Nursery stock. Specification for Irees and shrubs', <https://shop.bsigroup.com/ProductDetail/?pid=000000000000262241> [accessed 20 June 2019].

18 'HoloLens is helping architects to build better buildings, says RIBA', in *Microsoft News Centre UK*, 2017, <https://news.microsoft.com/en-gb/2017/07/20/microsoft-hololens-is-helping-architects-to-build-better-buildings-says-riba/> [accessed 15 March 2019].

19 'ASLA Code of Professional Ethics | asla.org', <https://www.asla.org/ContentDetail.aspx?id=4276> [accessed 7 March 2019].

20 'what3words | Addressing the world', <https://what3words.com> [accessed 20 June 2019].

21 'United Nations Disaster Reporting App & what3words', in *what3words*, <https://what3words.com/partner/un-asign/> [accessed 20 June 2019].

22 D Mitchell, 'We're unable to see the wood for all the pesky bird-filled trees', *Guardian*, 17 March 2019, <https://www.theguardian.com/commentisfree/2019/mar/17/unable-to-see-the-wood-for-all-those-pesky-bird-filled-trees> [accessed 20 June 2019].

23 'Disturbed by noise in the Square Mile?' City of London, <https://www.cityoflondon.gov.uk/business/environmental-health/environmental-protection/Pages/Disturbed-by-noise.aspx> [accessed 11 April 2019].

24 'Scheme Monitors | ccscheme', <https://www.ccscheme.org.uk/ccs-ltd/site-monitors/> [accessed 20 June 2019].

Chapter 6

1 Based on the Landscape Institute Digital Plan of Works for Landscape – Release 1.0, created by Anna Dekker and the LI BIM Working Group for The Landscape Institute, 2017 <https://www.landscapeinstitute.org/technical/bim-working-group/li-digital-plan-of-works-for-landscape/> [accessed 18 July 2019 and licensed under CC BY 4.0].

2 'Management and maintenance plan guidance for landscapes, parks and gardens | The National Lottery Heritage Fund', <https://www.heritagefund.org.uk/publications/management-and-maintenance-plan-guidance-landscapes-parks-and-gardens> [accessed 20 June 2019].

3 M Crawford, *Creating a Forest Garden: Working with Nature to Grow Edible Crops*, Reprinted with minor amendments; Hier auch später erschienene, unveränderte Nachdrucke, Totnes, Devon, Green Books, 2012.

4 L Kolowich, '22 Customer Review Sites for Collecting Business & Product Reviews', <https://blog.hubspot.com/service/customer-review-sites> [accessed 8 April 2019].

5 A Henry, 'Get Better at Getting Better: The Kaizen Productivity Philosophy', in *Lifehacker*, <https://lifehacker.com/get-better-at-getting-better-the-kaizen-productivity-p-1672205148> [accessed 11 April 2019].

6 D Clark, 'What You Need to Stand Out in a Noisy World', in *Harvard Business Review*, 2017, <https://hbr.org/2017/01/what-you-need-to-stand-out-in-a-noisy-world> [accessed 20 June 2019].

7 'CPD | Landscape Institute', <https://www.landscapeinstitute.org/member-content/cpd/> [accessed 25 March 2019].

8 'The Future State of Landscape Architecture – Embracing the Opportunity', 5.

9 *The Green Book – Appraisal and Evaluation in Central Government*, 2018, <https://assets.publishing.service.gov.uk/government/uploads/system/uploads/attachment_data/file/685903/The_Green_Book.pdf>.

10 *The Magenta Book – Guidance for Evaluation*, 2011.

11 M Nourinejad, S Bahrami & MJ Roorda, 'Designing parking facilities for autonomous vehicles', in *Transportation Research Part B: Methodological*, vol. 109, 2018, 110–127.

12 Ibid.

13 C Werth, 'How Stupid Is Our Obsession With Lawns? (Ep. 289), *Freakonomics*, <http://freakonomics.com/podcast/how-stupid-obsession-lawns/> [accessed 26 April 2019].

14 FS Santamour, 'Trees for Urban Planting: Diversity, Uniformity, and Common Sense'.

15 T McVeigh, 'Dieback has affected 90% of Denmark's ash trees. Britain faces a similar threat', *The Observer*, 6 October 2012, <https://www.theguardian.com/world/2012/oct/07/disease-killing-denmarks-ash-trees> [accessed 26 April 2019].

16 Santamour, 10.

17 'Level Monitoring', in *Flood Network*, <https://flood.network/what> [accessed 20 June 2019].

18 E Hubbard, *John North Willys – Elbert Hubbard's Selected Writings*, Part 2, 1922, 335.

19 'Latent Damage Act 1986', <http://www.legislation.gov.uk/ukpga/1986/37?view=extent> [accessed 27 April 2019].

20 'Learning Legacy | London 2012', <https://webarchive.nationalarchives.gov.uk/20180426101359/http://learninglegacy.independent.gov.uk/> [accessed 28 April 2019].

21 An evaluation approach introduced to me by the monitoring and evaluations manager from the Stockholm Environment Institute, Annemarieke de Bruin, that combines questions from S. Earl's *Outcome Mapping*, R Davies and J Dart's *Most Significant Change* and L Meagher and C Lyall's work on academic research evaluations.

22 'Landscape Institute archive', *The MERL*, 2017, <https://merl.reading.ac.uk/collections/landscape-institute/> [accessed 20 June 2019].

23 'File formats and standards – Digital Preservation Handbook', <https://dpconline.org/handbook/technical-solutions-and-tools/file-formats-and-standards> [accessed 28 April 2019].

24 'Recommended Formats Statement – table of contents | Resources (Preservation, Library of Congress)', <http://www.loc.gov/preservation/resources/rfs/TOC.html> [accessed 28 April 2019].

25 *Public Lab: a DIY environmental science community*, <https://publiclab.org/> [accessed 20 June 2019].

26 University of Bristol, '2013: Sustainable landscapes for the future | Cabot Institute for the Environment | University of Bristol', <http://www.bristol.ac.uk/cabot/events/2013/298.html> [accessed 25 April 2019].

27 H Wood, 'Sustainable Landscape Management – from past experience to future perspectives', presented at the Landscape Institute North East Region, Royal Station Hotel, Neville Street, Newcastle upon Tyne, 2016.

Appendix

1   IFLA Europe Code of Ethics and Professional Conduct.

IMAGE CREDITS

Architectural Press Archive / RIBA Collections Figure 6.16

Ben Ward Figure 6.11

The Broads Authority Figures 2.0, 2.2.0, 2.2.2–5, 2.2.7–8

Centre for the Protection of National Infrastructure Figure 4.1.0, 4.1.6, 4.1.9

The City of London Figure 5.10

Considerate Constructors Scheme Figure 5.11

Claire Thirlwall Figures 1.0–1, 1.3, 1.6–7, 2.1–2, 2.4–5, 2.7–10, 2.2.1 2.2.6, 2.2.9, 3.0, 3.7–10, 4.0–3, 4.5–14, 4.1.1, 4.1.3–5, 4.1.7–8, 4.2.0–1, 4.2.4–5, 5.1–5, 5.7–8, 5.1.3, 5.1.5, 5.1.7, 5.1.9, 6.0–6.2, 6.5–10, 6.13, 6.17–18

David Jarvis Associates Figures 1.5, 3.5, 6.14–15

Emmanuelle Martos for Ilex Paysage + Urbanisme Landscape Architects Figure 6.1.4

Gustafson Porter + Bowman Figures 5.1.0–1, 5.1.4, 5.1.6, 5.1.8, 5.1.10

Hani Hatami, Humanscale Figure 4.4

Howard Wood Figures 2.3, 6.1.0–3, 6.1.5–8

International Living Future Institute Figures 1.1.2–3

Jacqueline Cross Figure 5.0

James Hitchmough Figures 4.2.2–3, 4.2.6–9

Jeff Schmaltz, MODIS Rapid Response Team, NASA/GSFC NASA Figure 5.6

Jeremy Barrell Figure 6.3

Landscape Institute Figure 6.4

Landscape Institute Archive Figure 6.12

Maggie's Centres Figure 2.1.2

Maggie's Centres / Arabella Lennox-Boyd Figure 2.1.0

Maggie's Centres / Charles Jencks Figure 2.1.8

Maggie's Centres / Lily Jencks Figures 2.1.3, 2.1.6–7

Maggie's Centres / Rankin Fraser Figures 2.1.1, 2.1.4, 2.1.9

Maggie's Centres / Rupert Muldoon Figure 2.1.5

Mark Farmer Figure 2.6

Martin Brown Figures 1.1.7–8

Morgan Sindall Construction & Infrastructure Ltd Figure 5.12

Paul G. Wiegman, courtesy of Phipps Conservatory and Botanical Gardens Figures 1.1.0–1, 1.1.4–6

RIBA Collections Figures 1.2, 3.1–4, 3.6, 4.1.2, 5.1.2

Talley Associates Figure 1.4

what3words Ltd Figure 5.9

INDEX